Widdershins

Widdershins

Stories of a
Fisherman's Daughter

by

Nellie P. Strowbridge

Published by
Jesperson Publishing Limited

Jesperson Publishing Limited
39 James Lane
St. John's, NF Canada
A1E 3H3

Cover and Book Design: Donna Snelgrove
Printing and Binding: Jesperson Press Limited
Cover picture: A serigraph (Limited Edition)
Canadian Artist Gary Kennedy
Fisherman's Daughter

Jesperson Publishing acknowledges the financial support
of the *Canada Council* towards its publishing program.

The publisher acknowledges a financial contribution from the *Cultural Affairs Division* of the *Department of Culture, Recreation and Youth* of the Government of Newfoundland and Labrador which has helped make this publication possible.

Widdershins has some authenticity in settings and lifestyle, and the characters have elements of *roman-à-clef.* However, the book is fictional.

Printed in Canada.

Canadian Cataloguing in Publication Data

Strowbridge, Nellie P., 1947-

   Widdershins

   ISBN 0-921692-73-0

I. Title

PS8587.R68W43 1996   C813'.54   C96-950092-0
PR9199.3.S77W43 1996

Dedicated to —
my parents: Clayton and Lillian (nee Upshall) Kennedy,
my brothers: John, Jacob, Charles, George, Gary,
and sisters: Betty, Carol Ann and in memory of Nancy.
To my husband, Clarence Strowbridge,
daughter, Janalee and son, Michael—and to all the people
whose genes are from the same genetic pool as mine:
Kennedys, Upshalls, Taylors, Maleys, Peddles, Dicks,
Busseys, Pettens, Porters, Dawes, Ralphs,
Battens, Greeleys, Lears and Morgans.
Also, for all those baby boomers who would like
to turn back the clock—do Widdershins—
for a last idyllic look at days that were.

Contents

Acknowledgements

To The Newfoundland and Labrador Arts Council for its financial support.

To Robinson-Blackmore Printing and Publishing Limited for publishing **Old Nart** simultaneously in *The Compass*, *The St. John's Metro*, and *The Southern Gazette* (First Rights Only).

To adjudicators—Tom Dawe, Elizabeth McGrath, and Helen Porter—for their recognition of my work in "The Newfoundland Arts and Letters Competition."

She was almost glad when she slipped and fell. Down she sank into a warm, feathery bed, feeling something like a skinned bird might feel at finding its feathers again. She closed her eyes and drifted with the snow....

OLD NART

It was almost Christmas, and the only present Mandy wanted was a doll. She pictured her eyes, shining like marbles, making a plopping sound inside her head as her eyelids, fringed by soft eyelashes, closed over them. It wasn't that she didn't miss her Annie doll, who had always slept with her eyes open because they were painted on. But Annie was dead. She had swelled and cracked like a cake of hard bread in soak, after Mandy's brother, Jeff, threw her into a puddle of dirty brown water.

Mandy knew her father wasn't going to buy her a doll. He hadn't received a cent of unemployment money since he pulled up the boat last fall. A man down the coast, who was also Eric Maley, was getting her father's cheques, and there wasn't a sound of help coming from those in charge of unemployment payments. She couldn't depend on Santa either. Her father said that Santa Claus was some pagan invented by rich merchants to spoil children with promises their poor parents had to try and keep. Getting a doll was up to herself.

And this was all that Mandy could think about as she and her ten-year-old friend, Amy, went sliding on Big Hill in the Cove.

The hill was like a slippery white boot over which they could race right to the bottom, then coast along the foot to the edge of the cliff, which happened to be right beside Old Nart's door. The old man had returned to the cove a few months earlier, after spending most of his life in Boston. There were whispers about him and voices like walls discouraging the children of the cove from going near him.

In a snow bank, just outside his grey, two-storey house Mandy spied the money. She reached for it, but Amy was quicker. She grabbed it, then took off her cuffs so she could smooth out the bill. "There's number one and two zeros on it," she gasped,—"a hundred-dollar bill. I'm rich!"

"I saw it first," said Mandy, reaching for it again.

"You could see it all day," reasoned Amy, "but unless you got it in your hands, it's not yours." Then she pulled her cuff on over the money and grabbed the rope on her slide. Mandy watched as Amy started running. The long woollen coat that her mother had turned to hide its shabbiness slopped against her white fur top boots.

An ache stirred in Mandy. The money could buy her a doll. "You'll have some Christmas," she called after Amy, "once your father gets his hands on the money. You'll be rolling across the floor on beer bottles, and your father's legs gone to jelly. He won't be able to stand long enough to cut you a Christmas tree."

"My father won't see it." She kept on running and taunting, "Getters, keepers!"

Mandy went home and dropped down on her bed. She was thinking about the money when through the window she saw Amy running across the icy road. She jumped up just as Amy stopped below the window and called up, "The money, Mandy; it's funny money."

"Shush," said Mandy, pushing up her window, and sticking

her head out. "Open your hand for goodness sake. The money won't fly away."

"See," said Amy breathlessly. "Queen Elizabeth is missing. Instead there's a man on it with woolly white hair—looking as old as the hills. If I take it down to the shop, Aunt Sara won't give me anything for it."

"That's what you get for being greedy," Mandy said with satisfaction. "That's George Washington, the first American president—or," she peered uncertainly, "maybe it's Benjamin Franklin. Never mind, it's money." Then it dawned on her that the money could belong to Old Nart. Mandy had heard Aunt Sara talk about Old Nart coming into her shop and peeling a bill off a roll of American money to buy cheese and bologna. She said he had real old coins too.

Aunt Edith, the oldest and sternest-looking woman in the cove, was often at the shop chawing about someone. She said Old Nart stole his brother's girl and took off to "The States." Now he was back all grown over by a beard that hung under a mop of steel wool hair. "If it weren't for his cats," she declared, "birds would build a nest right on top of his head—and he's mare-browed too."

"Mare-browed?" Mandy had ventured.

Aunt Edith had pushed a long finger under Mandy's chin and spoke in a stern voice, "That means his eyebrows meet, and anyone whose eyebrows meet is unlucky. Beside, he can put a bad spell on you."

When Mandy asked her mother about Old Nart she told her he was just a poor old fellow the blessed Saviour died for.

"Then he doesn't sleep on a grapnel?"

Her mother shook her head, "Pay no mind to the gossip trough. Some people have a tongue for prayers and one for *prating*, and the less they know about someone the more they got to say—especially if he's got the smell of a faraway place on him. That poor man's only sin is being smart enough to stay put until hunger drives him out for a bite."

Amy brought Mandy's thoughts back to the money. She reached up to pass her the crumpled bill. "Here," she said, "I'm going to wait for Santa to bring me a present." Amy acted as if Santa could get down her chimney, even if she didn't have a fireplace, that he could just let the air out of himself and squeeze through a crack or nail hole. Then he'd blow himself up again.

"You don't have to depend on Santa," Mandy insisted. "Stay right there." She ran across the hall into the cold front room. She pulled a small wooden box from a shelf in the bookcase. Her father kept the graveyard box. The money in it was donated by parishioners to fix up graves in the spring. She counted eight dollars and forty-eight cents. That should be enough for two dolls. Then she put the hundred-dollar bill into the empty box. She brushed aside the guilt that hit her in the chest as she hurried out of the room.

"Only that much," exclaimed Amy in dismay.

"It's more than we had when we went sliding," reasoned Mandy, "and Dad's going to Bayview to buy some twine and nails. We can hide in the house on the truck. While he's in the shop we can go to the drugstore and pick out our dolls."

When Amy wasn't greedy she was being a copycat. The doll Mandy chose was the one Amy wanted. Luckily, the doll had a twin sister sitting right beside her. Leftover change was plopped down on the counter for some sweets from a big round bottle.

Mandy let out a sigh of relief when they reached the back of the truck. "I'm hiding mine under the bed until Christmas," Amy whispered. "Then I'll tell Mom that Santa brought it. She won't know the difference."

The Sunday before Christmas came, and with it the usual prayers. Mandy sat in the mission church trying to get her thoughts of Old Nart to go away. As was often the case, she was present in body but not in spirit until the minister, with a loud stamp of his feet, as he was journeying through his sermon, brought her body and spirit back together. When he said that foolishness was bound up in a child, Mandy knew he was talking about her and her insides got heavier and heavier. She couldn't take the American money out of the box. Then it would be empty. She'd have to give Old Nart her doll.

Amy said she wasn't going to give up her doll. "If you give up yours," she told Mandy, "you'd better go as a mummer, so the old man won't know who you are. When he opens the door, you can throw the doll in."

It was Christmas Eve before Mandy could talk herself into taking the doll to Old Nart. She'd already given her a name, and now she wrapped Angel inside her coat and sneaked out. The wind clipped her ears and pushed at her back, making her go so fast her legs got almost too tired to carry her body. Between the cold and the fright of having to face Old Nart she was *bivvering*.

She knocked at the door and soon found herself facing a man she'd seen only from a distance. His eyebrows hung like mounds of snow over cliffs circling fierce-looking eyes, and she drew back.

"You be a mummer, I suppose," he snorted.

She pushed her bandanna back over her head and stammered, "No—Sir, I'm Mandy, and I'm cold." She could see in around the door where a brown Siamese cat sat glaring at her.

"Don't mind the cats, Girl. On winter nights they lie on me as pretty as a patchwork quilt, and keep me warm. I hope they give out before I do, else they'll likely make a raid on the old bones—in you go then, right over by the fireplace so you can dry that iron rust hair of yours."

"It's not iron rust and it's not wet," said Mandy in an injured tone. "It's copper, like a new cent."

"That it is then, and as long as a cat's tail."

Mandy looked around in astonishment, "It's clean in here." Her hand flew to her mouth. "Oh!"

"I see you've been hearing the rumours that have been flying like gulls." He took a pipe from the mantel and lit it. "I suppose," he said, "you know about Elena. They say I stole her from my brother. It took my brother his whole lifetime to realize that I stole only Elena's heart. The rest was up to her. When he died and left me the house I realized that not hearing the voices of the sea all these years I was as good as deaf. I may not belong to the people in the cove, but I belong to the cove, the same as the land and the sea."

Mandy remembered that she had something belonging to him, and her story about the money tumbled out. "It's Christmas tomorrow," she added, "and the new year is coming, and I can't go into the new year with the old year stuck in my conscience. Here, you can have my doll." She laid it on his arm and started for the door.

Old Nart caught hold of her hand and drew her back. He laid the doll back into her arms. "Dolls may keep little girls company," he said gently, "but an old man needs people. I didn't miss any money and what I don't miss I don't need. 'Tis a good thing you found it first." He chuckled and scratched his chin through a thicket of bushy hair. "If the wind had snatched it and taken it out to sea, the fish wouldn't have been as obliging as it was to Saint Peter."

Seeing that the old man was kind, Mandy reached into her pocket and took out a red tie. "My father never wore it," she explained. "It's a present. It doesn't look like you have any. Are you waiting for Christmas?"

"At my age, just being here is a present," he smiled.

"You can wear your present to church," Mandy said brightly.

He shook his head, "I don't believe in church."

"But you believe in life—your life and God's life. That's what it's about."

"The little bit of life that's in me isn't worth much, but I believe in something. I once saw someone bring down twenty-foot waves by making the sign of the cross."

Mandy looked at him doubtfully, "Waves always come down."

He took a puff on his pipe and squinted at her, "But they don't always stay down, not when there's a hearty storm on the go. Anyway, 'tis a fine name you got, Amanda, Latin for worthy of love, and virtuous in thought and deed. That you are, Girl, but you had better be running home now, as fast as your legs can carry you."

It was stormy outside. Snow pellets hit Mandy's face like hail stones. She was getting tired of trying to push her way through the gale. If only she could tie herself to a gust of wind and be home! She was almost glad when she slipped and fell. Down she sank into a warm, feathery bed, feeling something like a skinned bird might feel at finding its feathers again. She closed her eyes and drifted with the snow....

The sound of sleigh bells brought her up out of a deep comfort. She couldn't believe what she was seeing: Santa, all covered in white like an angel, was lifting her into his sleigh. She fell sound asleep, and woke up under a mound of bed clothes. She jumped at the sight of the bulging stocking at the foot of her bed. Christmas morning had come!

She was biting the points off her apple and sniffing its cold, sweet smell when her mother came into her room, and put her doll in her arms. "I know what mischief you've been up to, and we'll talk about it after prayers," she warned.

Mandy felt almost too tired to go to prayers Christmas Sunday morning, but as soon as she turned and saw Old Nart sitting right behind the pot-bellied stove she was glad her mother had made her come to church. The old man was wearing the red tie, and suddenly he wasn't just a grey and white picture. He blended in with everyone. The Maley Cove people were so surprised to see him that they all shook his hand as he was leaving.

Mandy caught hold of Old Nart's coat as he was going through the door. He turned, and she looked up at him breathlessly. "I didn't believe in Santa Claus, but I saw him last night."

Old Nart winked and a smile spread across Mandy's face. She winked right back.

...She was horrified to see a red stain on her brown ribbed stockings. Panic rose in her like an animal trying to claw its way up through her throat. The door was pushed against the wind. Unheeding the key that pressed into her palm like a sliver of ice, Mandy ran calling, "Aunt Callie—Aunt Callie!"

SECRETS

For years Mandy had known of the charm. At night she'd lie in bed recalling snatches of conversation she'd gathered from her mother's relatives when they thought she wasn't paying attention. Most of them spoke skeptically about Aunt Callie's ability to stop blood. "It's the devil's work," her mother once sniffed to Mandy's father. "Your sister should be stopped." And while her mother seemed content to go to her grave without ever learning Aunt Callie's secret, Mandy was eager to discover it.

The summer she turned eleven, Mandy could hardly hold back her excitement as she waited for school to end. Every year, for as long as she could remember, she'd spent two weeks with Aunt Callie and Uncle Gus. Her senses tingled with anticipation as her father helped her aboard their thirty-foot skiff for the twelve-mile trip to Smith's Point. This was the summer she must learn about the charm; she'd made up her mind to that.

Even before the boat reached the shore, Mandy could make out the stout figure of her aunt waiting at the edge of the steep bank. The figure got bigger and bigger as the boat knifed its way through the rippling waters. In spite of her father's warning to be careful, she hurried over the side of the boat as soon as it got close enough to the land. The black waters made sucking noises against the sharp rocks below as Mandy's rubber boots plopped her against the soft turfed landing. Without waiting for her father's help, she crawled up the grassy slope. She was afraid that if she straightened up she would fall into the blackness of the sea. Reaching the top of the embankment, she eagerly grabbed hold of a wooden rail, then swung herself and her small suitcase onto the gravelled road.

Her aunt stood waiting, rubbing her heavy hands down her white bibbed apron. "Glory be, Child," she exclaimed, "you've grown some, and as polished as a button, too."

"Mom made me scrub," Mandy answered absently. Growing, or looking like a button was not what was on her mind as she followed her aunt along the dusty road.

"Aunt Callie," she blurted out, "are you a witch?" The words leaped off her tongue almost without her thinking them. Perhaps it was the sight of the two-storey house that rose in the air and came to a point like the peak of a witch hat.

She watched anxiously as Aunt Callie's jaw dropped and her chin sagged and became two. She straightened her face to ask tolerantly, "Now Child, do I look like a witch?"

"No, I don't think so," Mandy answered, now unsure of herself. Witches, she knew, had long noses and even longer faces; her aunt's nose was short and fat and her bumpy cheeks were wrinkled like shrivelled potato skin. "But you must be almost a hundred years old and have lots of secrets," she insisted.

"I'm seventy, Child—seventy." Her aunt sounded annoyed. She stopped to push a clump of grey hair up inside her black cap, then moved along the road more briskly. Afraid that she had offended her, Mandy fell silent.

She scuffed her way up the lane and followed her aunt around the corner of the house. Aunt Callie lifted the latch on the back door and they went into the porch. Mandy, determined to please, pulled off her boots and walked across the bumpy canvas-covered floor in her stocking feet. Her aunt frowned. "Put your boots back on," she said sharply, "you'll catch your death."

Mandy quickly shoved her feet down into her boots and took her suitcase into the hall. Uncle Gus, from his oval frame on the wall at the top of the stairs, looked down at her. Dashingly handsome in his Navy uniform, this man seemed like a stranger, not the uncle she knew. Suddenly a ramshackle version of the man in the photograph came up from behind, grabbed her with his strong, beefy hands, and smacked a wet kiss on her face. The stench of his chewing tobacco made her nose feel pinched as if she was suffocating. Perhaps he'd forgotten that he was no longer the man in the picture, the young sailor who had courted her aunt on a train and proposed to her—all in one day. Mandy's mother was fond of declaring that the marriage had gotten off on the wrong track. "A woman chaser," was what Mandy had heard relatives say in low tones behind their hands. Now she wiped away the offending kiss, no longer feeling the urge to crawl under the table as she had done in other years when her uncle had grabbed her.

Later, as she lay in the big feather bed in the spare room, Mandy pulled the heavy quilts up around her face and tried hard to think of a way to get the secret of "the charm." As she lay in the darkness, all kinds of ideas came to her mind. But gradually, one by one, her thoughts folded themselves away into the drawers of her mind and she drifted into sleep.

Morning came so fast that it surprised her. She sat up quickly, rubbing her eyes against the brightness of the sun pouring through the window. The sweetness of lilacs and honeysuckle drifted across her nose. She took a moment to enjoy the feel of her toes against the tightly fitted bed sheets, then bounced out of bed and pulled the quilts up over the big feather pillow. She washed and dressed quickly, then bounced down the stairs straddling two steps at a time.

Her aunt was looking into a small mirror on the wall and buttering her face with Noxema. Mandy wetted her lips nervously. "Please, Aunt Callie," she asked, "won't you tell me how I can stop blood?"

"My gracious," her aunt declared, "You are a curious child." She turned and patted Mandy's hand with her large greased one. "Perhaps I'll tell you some day," she said.

"But when?" Mandy persisted.

"When you're a woman," her aunt answered with finality, going to the pantry to wash her face. "Now run down to the chicken coop and see if there's an egg for your breakfast."

"Eggs!" she shrieked, and took off like a bird, frightening two hens out of her way. She found one warm, brown egg lying in sawdust. She cupped it gently. The thrill of finding an egg for her breakfast stilled her curiosity about the charm. She hurried up the lane and into the house, where her aunt stood melting a lump of margarine in the frying pan.

While her aunt set about to make a batch of bread, Mandy sat down beside the oil-clothed table and ate ravenously. She was putting the last piece of egg up to her mouth when Aunt Callie interrupted her with a stern look. "Did you say Grace, Child?"

Mandy's eyes widened in alarm—an unconscious admission of guilt. Her mind had wandered back to the charm, and she'd forgotten everything else. She shook her head.

"And you've eaten almost every bite of food," her aunt added with an incredulous look. It was almost as if she expected her niece to choke for such a grave omission. "You'd better say it now," she told her. "I don't want your mother to think I've turned you into a heathen."

"Yes Aunt," Mandy answered in a meek voice. "Then will you tell me the charm?" she asked hopefully. "I can't bear to wait; it makes me so absentminded." Seeing the stern look in her aunt's eyes, she bowed her head and mumbled, "God, bless the food. Amen."

Her aunt stopped kneading the bread and turned a kinder eye in her niece's direction. "Your uncle is down in the barn," she said. "Go now, and there's a chance you'll see some new chicks."

Mandy began to despair that she'd ever find out her aunt's secret as she dragged her feet down the path to the bottom of the hill. In one corner of the barn, Mandy bent down to pick up a cracked brown egg. She picked at a piece of shell and a tiny bill poked through; she kept picking off shell until she released the tiny, golden chick. Crusted blood clung to its feathers, but cupped in her hand the little creature looked like a blob of sunshine. Uncle Gus came and stood above her. He eyed her keenly. "You came from an egg," he chuckled. His voice was like a needle puncturing pinholes in her spine. She stood up abruptl. Just then a squirt of tobacco hit the side of the barn. Threatened by some nameless fear she followed a shaft of sunlight to the outside.

"Did he?" her mother once asked, "ever try to take down your pants?"

With a sense of shame which she didn't understand, and puzzled at why her mother would ask her such a thing, Mandy was glad she could whisper, "No." But after that, Mandy knew her mother had secrets—tightly wrapped secrets that she couldn't even guess at. Even Aunt Callie had other secrets. Her breasts were gone—cut off and they'd never grow back. Now there were lots of handkerchiefs filling up the empty pockets of her bra. Once when her mother didn't know she was near, Mandy heard her remark that the handkerchiefs came in handy whenever Callie had a cold. "Even Gus makes use of them when he has a sniffle or two," she heard Aunt Callie tell her mother. To her surprise, her mother had lifted an eyebrow and laughed, though Mandy didn't know why that was funny—not when her aunt had "them" cut off because they had lumps that could have killed her.

Mandy ran over the green slopes of land, slashed through knee-high grass, and picked golden buttercups as she swished by them. Overhead, the clouds drifted light and fluffy. Babyland was high above the clouds—that's where her mother had said she came from. Mandy imagined herself adrift on a pink cloud waiting to be born, while her brothers drifted on fluffy blue ones. Some girls at school, Gloria and Susie, said they came from bogs. Dirty, mucky bogs, Mandy thought with distaste. Jamie, Susie's older brother, said he came from his mother's stomach. Of course nobody believed him. Wouldn't it be wonderful if she learned how to stop blood? Then she could stop Jamie's nosebleeds and everybody at school would think she had magical powers. So intent was she on her thoughts and aggravated by them she promptly fell into a thicket of stinging nettles growing up against the picket fence.

"My glory," said her aunt in vexation, a sigh coming up and out of her throat like a last breath. "No one ever fell into them nettles before. Knock wood." Her knuckles hit hard against the pantry shelf and a dozen jugs swung precariously. "Your uncle will get the scythe and cut the things down."

Her aunt's home remedies covered the sting but left the itch and Mandy was forced to concentrate on her pain. Finally to compensate for what her aunt called her misadventure, Aunt Callie set about to curl Mandy's long mane of hair which— according to Uncle Gus, was the colour of a robin's red breast. Mandy eyed her aunt with pity when she mentioned such a task. Her hair was as straight as Uncle Gus's wooden leg, and not even a charm could change that. However, Aunt Callie had her way and within an hour Mandy's head held a dozen wet ringlets—aided by a bar of Sunlight soap, and a package of her aunt's metal clips. When the water dried out, the soap and pins stayed and Mandy's hair was like twisted strands of rope. Later it provided an instant bubble bath in the galvanized tub in the back room.

Mandy liked to go across the field to the outhouse (for many reasons). When she lifted the key off the hook in the pantry and made her way across the top of the hill she imagined that she was going to her own little playhouse. The brightly painted toilet stood

26

upon the hill like a matchbox on end. Every spring it was repainted in the same dazzling orange colour. Her aunt had cautioned her about leaving it unlocked. Now she slid the key in the lock and turned the glass knob. Pretty yellow curtains hung at its two windows; the walls were papered, and the slashes of red flowers on the wallpaper looked as if they were forever blowing in the wind. A picture of Her Majesty, Queen Elizabeth, was hung above the toilet seat. Wafts of ammonia drifting up to her delicate paper nostrils from the generous sprinkling of lime under the toilet, did nothing to disturb the wide, majestic smile. When Mandy had asked her aunt why she had hung a picture of Queen Elizabeth above the toilet seat, Uncle Gus had butted in with a grin, "That's where she belongs—on her throne." He always called the toilet seat a throne, though Mandy couldn't see why. She took her place beneath the queen's picture and tore a page out of the winter catalogue. To soften it, she crushed it in her hand, disfiguring the pretty faces of girls in bright dresses. She was finishing with the paper when she noticed a rivulet of blood coursing down her white leg. Her legs were white because they had almost never been held under the eye of the hot sun due to her father's unmovable assumption that legs, unless they were on hens, horses or furniture, should be covered at all times. However, when she was picking berries on the hills, she sometimes rolled her stockings down around her ankles like two donuts and enjoyed, with abandonment, the warm satiny fingers of the wind brushing along her bare legs. Now she was horrified to see a red stain on her brown ribbed stockings. Panic rose in her like an animal trying to claw its way up through her throat. The door was pushed against the wind. Unheeding the key that pressed into her palm like a sliver of ice, Mandy ran calling, "Aunt Callie—Aunt Callie!"

Aunt Callie's head jerked towards Mandy as she came running through the doorway. But before she could say anything, Uncle Gus, who had been having his afternoon snooze, pulled himself up from the daybed and muttered, "What's the matter with the maid now?"

"Stop the blood," she pleaded. "Please stop the blood." The freckles on her white face stood out like brown polka dots. She could feel her eyes bulging. Somewhere inside her, where the veins ran through the nooks and crannies of her body there was a leak. Her blood had gotten out—and she couldn't stop it from flowing.

Uncle Gus's steely-winged eyebrows swung upward into a question mark; then his wicked brown eyes followed the bright red stains on her stocking. He shot a mouthful of tobacco—quid squirt—her aunt called it. It fell into a bowl on the table, and for once Aunt Callie didn't chide him. Her voice was brusque as she told Mandy, "I'll get you some cloths and pins."

"But Aunt," Mandy felt the tears forming at the backs of her eyes, "you've got to stop the blood." Her voice trembled.

"I can't stop nature, Child," she answered with unusual sharpness.

"Nature?" Mandy asked in bewilderment. A red colour stained her aunt's wrinkled neck and moved up to her face. She turned to Uncle Gus. "I think you've got a few odds and ends to take care of down in the barn," she said primly.

As he passed by Mandy, her uncle winked at her, then came back to whisper in her ear, something about stopping blood for nine months. Her childless aunt gave him a push towards the door.

"Come now, I'll make you some cloths." Her aunt nodded towards the living room. Mandy followed her in. "You're now a woman," she said in a soft awkward voice. Mandy saw only one significance in that statement.

"A woman!" she gasped, "Does that mean you'll tell me the charm?"

Her aunt went to a shelf and took down a large, brown book. It looked like the "Doctor's Book" her mother kept locked in an old trunk. Mandy wasn't allowed even a peek inside its rusty-looking covers.

Aunt Callie marked a page in the book, and passed it to her. "Now is the time for you to learn something more important than the charm," she said gravely. "Go upstairs, wash yourself, and change your clothes, then read the page I've turned down. I'll sew you some cloths right away."

Mandy took the heavy volume in her arms and climbed the stairs. As she washed and changed her clothes, using a handkerchief to soak the blood that didn't seem so bad after all, she heard the sound of the Singer sewing machine. She plopped herself across the feather bed and picked up the book. She was a woman now. What did her aunt mean by that? The book fell open and a piece of folded paper fell out. FOLK MEDICINES stood out in bold lines at the top of the page. Mandy stared at the words underneath. Stopping blood: blood can be stopped by certain persons who recite a secret prayer, she read. That was all. With that, her curiosity about the charm was suddenly cut adrift like a piece of flotsam. Now she turned frantically to find the secrets that her body had held and hidden from her all these years.

Summer sounds and scents drifted into her room. The muslin curtains blew across her face as if they were sachets of honeysuckle. Across the bay, a lone gull winged its way over the crinkled blue sea, arching itself in the sky, and letting out a wild cry. Her aunt came with soft steps, leaving behind cloth rags and pins. Mandy read on, unheeding anything but the book.

Suddenly she heard her uncle's voice, and his footsteps on the stairs. Words like "dirty old man" came unbidden to her lips as the first shreds of discernment began to take shape in her mind.

Little Mary's face flashed in front of her eyes. Little Mary from across the cove—only six—holding a candy bar in her hand on a hot summer's day. She'd whispered that she got it from the old man in the shed. Her eyes were wide and frightened, and her hand was all dirty from the melted chocolate.

UNCLE ERLKING

Bart was often there, just inside the door of the old shed, so close Mandy could almost feel his hot breath coming through the seams. She didn't mind so much when she could see him, but hearing him shuffling inside made her uneasy. Sometimes, as she hurried past the shed on her way to school, she thought she saw him at the cob-webbed window. The school stood high upon a hill, nestled among boulders, and overlooking the sea. To get to it, Mandy had to pass the shed and the double-boxed house that Bart and his brother, Libe, shared.

Sometimes in the spring when the air was warm and filled with an energy that made her dance along the road in her new shoes, she forgot about Bart. It was so good to be able to walk on clay-baked roads. They were likely as smooth as Heaven's golden streets. She'd pass the turn in the road without even seeing Bart. Suddenly there he was, standing by the open door. His peaky little face, under a mop of black hair, lifted towards her. His finger was crooked like a hairy carrot just pulled from the ground; lines across it *grainted* with dirt. He patted his grey shirt pocket and asked, "You want to see a dried fish doctor?"

Mandy had already seen one of those ugly things that resembled the grey *carpenters* her mother often swept off the damp concrete floor in the basement, except that fish doctors were orange-coloured and larger.

Mandy would just as soon go through an auger hole as cross to where Bart stood. She kept on walking.

"Come here," he called menacingly, "or I'll knock your insides out. Dang yer," he muttered. "Dang yers all. Der youngsters won't talk to an old man no matter what he gives 'em."

"You want strawberry candy den—to colour yer lips," he called hopefully. She looked back to see him rolling the candy between his fingers, the colour staining them. She turned away, and he called after her, "I 'ope the bully bagger gits yer."

Mandy didn't care about the bully bagger. He wasn't real, and her mother had never threatened her with him. The erlking was something like the bully bagger. Miss Upshall, the teacher, said he was an evil goblin who haunted the forests of Germany and lured children to destruction. Sometimes the erlking carried them off and put a spell on them so that when they came home they were mute. No one knew their secret. Mandy decided that Bart was the erlking of Maley's Cove, though she never told anyone why she thought that. She wanted to tell the teacher about him but she knew the other kids would laugh at her.

She found herself laughing at Bart when she told her mother about the fish doctor. Her mother said that the poor man was lonely and she shouldn't laugh at him. "You know what happened to the children in the Bible who laughed at Elisha; two she bears tore them—maybe they even ate them."

"I won't go inside the shed," said Mandy shaking her head.

"There's no need to," her mother answered quickly. "Just be civil and respect your elders. They don't have long to live."

Mandy didn't care how long they had to live, especially Bart and his brother Libe. They had already used up their three score and ten years. Now they were living on borrowed time. They'd never be able to pay it back in Heaven or Hell, since there was only eternity there—an endless stretch that had no notches in it to mark off time. She had nothing against Libe. He hardly went farther than the smoke in his chimney, and the only place she saw much of him was at the mission where he went for prayers twice on Sunday. For years, Missioners thought he would never go to the altar. He was like a cake of hard bread, but one Sunday morning he was found sobbing on his knees. As soft as brewis, he was. Mandy didn't mind saying hello to him if she passed him on her way to school. He'd lift his face, squint his eyes at her and grunt the o in hello.

There were jokes in the cove about the two brothers. They'd be sitting down beside the shed and Bart would ask, "Got der time?" Libe would take out his watch and answer, "Now I do," and he'd put the watch back into his pocket. When Bart would ask, "What does yer time say?" Libe would mutter, "Ticktock."

That's why people said Libe wouldn't give his brother the time of day. It seemed to be why Bart was always on the go, trying to find someone to talk to him. He'd save up all the figures he'd whittled from wood during the winter, then go around selling them during the summer months. He'd have them wrapped in blue derby, gingham, and London Smoke materials, and put with other things he'd scrounged from merchants for a few cents. The old peddlar would come down the road with a brin bag on his back, his arms hanging loose from his elbows, hoping to find people home so he could have a chat and pry some money loose from hands that were too busy making money to spend any. Sometimes Bart would find the women and children on the flakes spreading salt fish, or standing in fish slub on stageheads helping the men clear away fresh fish. Other times, he'd catch the women at home, some of them with their hands deep in bread dough. He'd knock, then push his way inside. The brin bag would be dropped from his slight shoulders to anywhere he could find a spot for it. He'd squat beside it and look into the face of anyone who was near. Sometimes it was a child who had followed him down the road just to get a molasses knob or humbug candy. If one was bawling or hiding behind her mother and sucking on the hem of her dress, he'd be sure to charm her with a sweet.

It would take more than a candy to charm Mandy. Even though she was twelve, she could not look into his piercing, black eyes without feeling fear. She wondered why he had a TD pipe upside down in a mouth that looked all grown over because of the straggly black beard that fell to his Adam's apple. It matched the straight black hair that was all over his head. He also had a black spot on his face that looked as if a spot of tar had dropped there and dried, then cracked. There was a matching black spot on his hand. Mandy noticed it when Bart's bony hand reached to open the brin bag. She wondered if the two black spots were patches of devil skin.

Her mother usually bought something just to satisfy him: a piece of material for a dress or shirt, perhaps a buckle for Mandy's bunch of unruly red hair. She always called him Uncle Bart out of respect for her elders. (Most of the cove people called older people—and people who looked older than themselves—Uncle or Aunt even if they were no relation.)

Some people were nice to Bart's face only because it was the Christian thing to do. They didn't think they should invite him to eat with them. Mandy's mother often asked him to stay for a cup of tea, and he always accepted with a grin that showed his few teeth, the colour of corn kernels rotting on the cob. If anyone suggested he get false teeth, he'd snort, "I'd radder 'ave 'arse shoes in me mout'."

Mandy watched as he poured tea from his cup into his saucer and dipped his lassy loaf in it. The soggy bread would drip into his beard. He'd wipe it dry with his sleeve. Sometimes he'd look across at Mandy and keep looking at her until she became so self-conscious she couldn't keep her head up. It would drop like a *piss-*

a-bed whose stalk had gone limp. When she thought he wasn't looking at her she'd slowly hoist her head.

She wished he'd go. Instead he stayed *mimpsing* a second cup of tea from a brand he'd probably just sold her mother. He believed King Cole was king of England. Mandy watched him, after he'd finished a meal of herring, run his fingers through his hair, leaving herring bones looking like short, grey hairs. It was no wonder some people didn't want him at their table. But Mandy's mother insisted that he was lonely, and a person had to overlook his defects in hopes of having him clean up his soul. He'd be clean, she figured, if he had a woman to look out for him.

Bart was going off to the post office quite often lately. He looked as if he was expecting something wonderful. If Mandy saw him ahead of her on the road she'd slow down. If he was behind her she'd speed up. One day her heel was galled by her new shoes and she couldn't hurry. He caught up with her. "'ello me maid," he said in a jaunty voice, "'tis sum' marnin' anin' it?" He patted his pocket and laughed slyly, "I got me notice—I'll 'ave me woman soon." He hurried on past her, looking quite harmless. Perhaps if there weren't so many things said about him she wouldn't feel so frightened around him. Like Gran Maley said, looking over the top of her wire-rimmed glasses as she sat knitting a pair of cuffs, "Everything you hear you got to turn it over on your tongue, turn it over some more, then chew on it before you spit it out into someone's ear. It could be filthy lies. And if you spread lies you're telling them, and got yourself swinging on the devil's tail."

When Mandy squeezed in among the crowd in the post office, she could hear the slapping sounds of the sea against the beams that held up the end of the post office jutting out over the water. People grumbled that the sea was likely to push the legs right out from under the post office. Behind the closed wicket were the sounds of the post mistress sorting the mail. She was called Aunt Hilda, though she wasn't anybody's aunt. She seemed to belong

35

to everyone in the cove. Mandy could see Bart close to the wicket, his neck stretched up. He was probably on tippy toes.

Suddenly the wicket slid back and Aunt Hilda's ample bosom lifted towards them. She was still puffing from lifting heavy packages and she had to catch her breath before she could start calling names. There was a shift in the packed room as names were called. Voices answered and mail was passed from hand to hand until it reached the people owning it. Aunt Hilda scanned the crowd and told this one and that one there was no mail. Disappointed, they turned towards the door loosening the crowd as they went out. Someone else moved closer to the wicket. A name was called and someone answered back, "I'll take hers." A parcel was passed over heads until it reached the right hand.

Mandy waited expectantly. Aunt Hilda's eyes had pitched on her, and she hadn't said, "Nothing for the Maleys." There must be something, she thought, even if it was some free stamps she'd ordered from *The Family Herald*.

She saw Bart when he got the letter. He lifted it above his head and looked up at it. His eyes looked shocked as they stuck into the letter. They fair bulged. You could knock them off with a stick.

Knowing he couldn't read or write, Aunt Hilda called, "Uncle Bart, the red dress you ordered is out of stock."

His eyes swivelled towards her. "Dress," he muttered, "I ordered a woman—a pretty one, too."

"Then she's out of stock," Aunt Hilda said, shaking her head at his stupidity. There was a sudden silence. Everyone moved out of the way to let him out the door.

There was such a look of misery on Bart's face that no one laughed for fear of cutting him. Later, those who hadn't seen him when he got the news laughed hard enough for everyone, and told about it—adding on—or, as Gran Maley would say, "Knitting sleeves in socks."

As Mandy left the post office, Uncle Tom stopped to pick her up. He was one old man that Mandy didn't mind calling Uncle even if he wasn't her uncle. More than once on her way from the post office he'd stop his horse and cart and reach out a heavy hand to lift her up beside him, saying, "Fine day for a trot." He'd pull the piece of brin bag from under him and let her sit on it on top of the pile of clay he was carrying to fill in the pot holes. She didn't mind getting clay on her brown ribbed stockings because it didn't show, but if she got her dress dirty her mother would call her a *ragmoll* or a dirty *streel* for messing a dress she'd just starched and ironed.

"You got *The Family Herald*," he'd say. "'Tis wonderful readin' idn't it?" He'd turn his eyes, the colour of beach rocks, towards her and she'd smile. Then she'd think about her grandfather, the one she didn't have, the one she always missed. She'd seen him only in snapshots, big and strong, with a face that smiled all over under his Sunday-go-to-meeting hat. Uncle Tom wore a hat, but it was to keep the hot sun in check. And he wore boots that smelled of oil and tar where he'd water-proofed them. Sometimes he polished his boots with Neat's-foot oil that he bought from Bart. But Uncle Tom was nothing like Bart or Libe, or the old minister who often visited Mandy's home before he went off to "The States." He would grab her and slobber her with kisses—unless she managed to skitter under the table. She thought it odd that he could bawl out the word of God in church so sternly, and then smile in a silly way.

It was a March morning, and Mandy was late for school. There was no one else on the road as she forced herself against the strong easterly wind. Spikes of hail stabbed her face, making it smart all over. She decided to cut across the pond instead of going around it. There was a lot of water on parts of it, and, with the wind gusting around her, she could hardly keep her balance. She wasn't wearing her leggings, just her navy fleece-lined bloomers. Her knees, inside her stockings, felt like junks of ice. A gust of wind swirled around her, bending her backwards and forwards on the slippery ice. She lost her balance and fell into the water. Every time she tried to stand, the wind pushed her down, edging her towards a black opening. She was almost to it when she noticed a rock stuck in the ice. She grabbed hold of it, and with all her strength she pushed herself away from the hole. Her knees felt rubbery and her drawers heavy and weighed down, as if they were being pulled off as she crawled ashore.

The sight of smoke snaking its way out of the chimney of the old shed on the hill made her forget her fear as she ran towards its comfort. She lifted the latch on the door and went inside. She dropped her book bag on a pile of wood stacked in a corner, and went towards the oil drum stove. Her whole body was shivering uncontrollably as she pulled off her water-logged boots and stood rubbing her feet dry on the shavings curled all over the floor. As Mandy warmed her hands on the the stove, her eyes stared at the cob-webbed window with dead flies and a live black spider. "Come into my parlour," said the spider to the fly. The words from a Reader rhymed through her head, and a different kind of shiver went through her heart.

She caught sight of the pictures of The Good Shepherd and Premier Joey Smallwood hanging side by side. Libe had a falling out with Joey and had moved his picture from the house to the shed. Bart said if Joey went, Jesus had to do likewise. Bart had told this to Mandy's mother when he tried to sell her a mirror with Joey on the other side, and a caption, "Will the lady on the other side please vote Liberal?"

Mandy turned quickly as the door swung open. She was afraid it might be Bart. It was only Libe. He looked at her with dull, water-galled eyes. His bottom lip stuck out like the mouth of a jug. "Came in fer a draw of breat', did yer? Dat win' 'ill suck it right outta yer."

"I fell down and got soaked," she answered tremulously, feeling sorry for herself.

Libe pulled on the light hanging from a nail in the rafters. "Soaked, I should say yer soaked—right down to der skin—good ting der skin is waterproof. You'd be drowned. No one else wit' yer?" He looked towards the door.

"No,"Mandy's voice was low, "they all left for school before me. Some got rides."

"Too bad," he smiled, his lips stretching like two water-plimmed worms."

"I'll—be late for school," she said haltingly. "Anyway I'm not shivering so much now."

Libe backed against the door. "You better git outta yer duds. I'll giv' yer me shirt. It'll fit like a dress and 'tis warm."

She looked towards his plaid shirt. She hesitated, then said, "Leave it here, then."

He moved closer. She stiffened. Suddenly it was as if there was another person looking at her from the backs of his eyes, someone she didn't want to know.

"You big *wis*," he laughed, "yer not afraid of me is yer?"

"No," she said quickly, "Bart—"

"Bart!" He shook his bald head, and laughed. "Why—dat gadabout is just capped wind. He's as 'armless as his dried fish doctor."

Mandy wet her lips and started to explain.

He cut in, "Yer actin' as if yer afraid of me."

"No!" She tried to keep her voice from quavering—tried to convince herself. "You're the good one. You wouldn't cleave a junk of wood on Sunday, even if you were freezing." She rattled on, telling him all the good things she'd heard about him.

"That's right, Maid," he said in an undertone, "God wants people doin' good things—like havin' fun wit da gadgets he gave 'em. Git der clothes off yer. Sure, steam is already drifting." He pulled off his plaid shirt, and Mandy blushed at the sight of the top half of his buttoned-up grey underwear.

Suddenly his voice was like a gimlet boring through her head, "This will cover ya, honey."

"No, I couldn't! And don't call me honey." Her voice shook indignantly as she moved towards the door.

"You know," he said in a coaxing voice, "wearin' clothes is our way of pretendin' we wasn't born naked. We're hypocrites. When Adam and Eve was really good dey wasn't wearin' clothes." He winked. "If God wanted us to be covered he'd 'ave put fur on us like he done wit' animals." His arm came down around her shoulder.

She pulled away and said evenly, "Jesus said not to offend little children."

He ignored her words and kept blocking her way. His voice was threatening. "Yer won't git over me time, Girl. I'm strong enough to pull der tail outta a bull." The two hairy hands that cupped her face like a vise made her shiver. Bart had the black spot, but Libe was the devil.

"I've got a bar right 'ere—in me pocket." She followed the bulge. He moved the bar out, then in—and then held it up like bait.

Little Mary's face flashed in front of her eyes. Little Mary from across the cove—only six—holding the candy bar in her hand on a hot summer's day. She'd whispered that she got it from the old man in the shed. Her eyes were wide and frightened, and her hand was all dirty from the melted chocolate. She had tried to get Mandy to take it. She said it was a dirty bar. Mandy had thought she meant Bart.

Libe's lips latched onto Mandy's. Her cry was barred inside her as she pushed hard against him. He tripped over a board and went down like a sack of potatoes. She just stared as blood oozed from a hole in the side of his head. Then she noticed a nail sticking up in a floor plank. Her shivering stopped, and she stood frozen by another fear.

The door opened, and she turned to see Bart. He caught sight of Libe. "What in tarnation!" He let out a rhyme of oaths.

"I—I -," she started stuttering, her whole body shaking.

"Never mind talkin'," said Bart, as if stupefied. "Git on wit' yer over to der school."

She ran out into the sleet storm. She ran without feeling the cold slush under her bare feet. "I killed him," she screamed so hard she felt as if her cry would pull her heart right up through her throat and drop it beating in the snow for Libe to get hold of.

She banged against the school door, grabbing hold of the knob, and looking hastily behind her, as she pulled the door open and stumbled inside. She stopped to catch her breath before knocking on the classroom door. When she did knock, she heard the teacher interrupt the lesson to tell a pupil to answer the door. Everyone looked at her with open mouths when she came inside, her teeth chattering so hard she couldn't speak.

"You're soaked!" exclaimed the teacher. "What in the world happened to you?" Then without waiting for an answer, she took her by the arm and led her over to the floor duct. "Stand here," she said kindly, "while I turn up the thermostat."

Mandy tried to push her words past the lump in her throat. Her dazed eyes lifted to the teacher's. "Miss," she said weakly, "we have Uncle Erlking here."

"Erlking?" The teacher's eyebrows came close together.

"Yes," her voice quavered, "it's Libe—and I killed him."

All eyes turned to Mandy. Some had a strange look in them. Even the teacher seemed to tremble, as if a shock went through her. She looked at all the children and spoke in a tight voice, "We won't talk about this now." She looked towards Mandy's older brother and said quietly, "Jeff, please go down to Mrs. Janes and call the parents. Tell them to come for the children."

There was a long silence, a silence broken finally by the sound of Mandy's father's footsteps as he came for her. He followed Miss Upshall into the cloakroom. When he came out he looked stern. Without a word, he picked Mandy up in his arms and carried her to the truck, laying her gently against the seat. He still didn't speak as he drove home, nor as he lifted her out of the truck and carried her straight to her room. He laid her on her bed and kissed her forehead. Then he went out and her mother came in. Her eyes looked puffed. She said, in a strained voice, "You don't have to talk about what happened. You can forget about it. It wasn't your fault. You better get some dry clothes on before you catch your death." She went out, closing the door so softly that Mandy felt that she was part of something so terrible everything outside it had to be kept quiet. There were no sounds from her three brothers, not even from little Timmy. The whole house had fallen silent, as if one of them had died.

She got up and went to the window. It had turned off freezing, and the window was covered in silver etchings of ferns and horses' manes. She put her mouth to the glass, melting a magical hole to take her back to the old one-room schoolhouse down by Aunt Lene's house.

Mandy loved to sit on the woodbox in Aunt Lene's porch before she went to school. The toe of her shoe would trace the bright red flowers, open like big cabbages, on the canvas floor, while she waited for Aunt Lene to put her dipper into a bucket of sparkling spring water and tip it to Mandy's lips.

Some people called Aunt Lene a good Christian soul. Her hair was always stuffed inside a black tam that cut across her eyebrows. She had a long bumpy nose and her chin stretched out from her face and rounded into a little round bun. She'd lost her teeth and her lips had foundered, causing her voice to slip out between her gums, soft and squeaky, "Come in, me babby."

Mandy knew Aunt Lene loved children. Her soft brown eyes would look at them, and she'd murmur, "After the children are here in the morning, I can hear their sounds all day—if I puts me mind to it."

Mandy had spent many summer hours lifting jellyfish out of the sea, her ten fingers like knives cutting them into jagged pieces before they fell back into the water. The jellyfish gave her warts, and Aunt Lene "put away" the fleshly barnacles.

But Aunt Lene was dead. Mandy's safe place was gone. Her fingers hit the glass. They scraped up and down the window. Silver shavings fell, marring the etchings and letting in the bitter look of the outdoors. She turned away from the scarred etchings and stripped off her clothes. She pulled on her flannelette nightdress and climbed into bed, pushing her feet down deep under the bed clothes until they were so warm and cosy that they felt crowned. Soon she was asleep.

When she awoke, the silver etchings were back as if they'd never been touched. There was no sign they had even been damaged. She caught sight of herself in the mirror. She looked the same too. All she had to do was pretend everything was the same, that nothing had changed. She looked into the mirror and smiled. She tried to push the smile up to her blue eyes. Trying as hard as she could, she could not make it reach. Instead, a tear leaked out of one eye and rolled towards her smiling mouth.

Pretending must be something grown-ups can do well, she thought—trying again.

She loved him. Her feelings rose to a fullness she could hardly take, and she couldn't bear the thought of him slipping farther away every moment that passed. There was an urgency inside her to see him, to touch him—to say goodbye.

ON THE WINGS OF AN EAGLE

Mandy didn't understand why she missed her Grandfather Maley. He was someone she had never known. She thought about him a lot, as she read a copy of his *Reader's Digest* she had found in her grandmother's shed. His eyes had been in the same spot hers were, reading the same words. Her mother said they had just missed each other in passing—Mandy into this world and her grandfather into the next. He died in March; Mandy was born in June. She imagined herself inside her mother's body listening to her grandfather's voice. On his way to Sunday morning prayers he often brought *coopy* eggs for her brother. Tall, big, sandy-haired and gentle, was how her mother described him. He died long before his three score years and ten.

That's why she liked Issie. He was someone else's grandfather, but he was alive. She couldn't wait to go next door to visit him. She was sitting at Aunt Callie's table watching through the window as her father's boat shrank to a speck in the bay and then disappeared. He had brought her to Smith's Point for her summer holidays. Now she'd be able to finish a book she'd started reading last Christmas, a book her cousin, Roy, had left at her aunt's when he finished university. It was called *Great Expectations* and was about an escaped convict and a little boy named Pip who went looking for a file to help the convict get rid of a leg chain. She hoped Roy hadn't come back for that book, or the others lying in a big box under the stairs. She was itching to read them all. She intended to have a good two weeks sitting out on the veranda smelling honeysuckle and letting the big eye of gooseberries burst inside her mouth. Some of the books were hard to understand. Sometimes she'd try and it was as if her mind would stretch and hurt to accommodate passages, but that didn't stop her. She knew that as time went on, she would understand more. She wished Issie could read so she could discuss books with him, but he couldn't read or write and he said he didn't care. "Years ago people lived and died believing the earth was flat, and not finding out never hurt or killed them," he reasoned. "There's only so much living you can do, and there's contentment in doing that well."

"I had a dream about Issie last night," said her aunt, lifting her eyes from her plate to look at Uncle Gus. "I saw him laid out on the old matchboard door that's hove across the drain outside his house. He had black flowers in his hands. Remember the time we heard the buzzing in the wall just before Mother died, and she up in Goose Cove, not near enough for us to know she was even sick? Then when we went there, her breathing made the same buzz. 'Tis an omen, for sure. He's not long for this world."

Mandy's mouth was full of mashed potato, and she couldn't speak. It had never occurred to her that the man next door would not be there as long as she lived. The first time she'd seen him she was about six. She had felt like Gulliver must have felt among the giants when Issie, standing in big leather boots, wiggled his finger in front of her face and said, "I'll sniff you up my nose." Fear rippled under her skin and popped goose bumps, but the twinkle that mysteriously surfaced in his eyes, made her smile. His silvery hair, over his head like a wad of tow, was the colour of the new nickel in her skirt pocket. His face was smooth and pink, making him fresh-looking, unlike someone who would belong in Bertha's kitchen. Mandy had seen Bertha slide a knife down her dirty apron before cutting bread, and cough over the pan she was frying Issie's breakfast of bologna and eggs in.

"Don't heave your tea out in your saucer," Bertha had scolded him.

"I shouldn't have one then." He had winked at Mandy, and dipped his bread in his *milquetoast* tea. He put it dripping to his lips, then remarked, "The Lord, himself, sopped his bread."

Issie's son, Bertha's husband, had died after he had caught 'ammonia.' That's what Bertha said. Mandy asked no questions, but she knew that her mother had ammonia and she hadn't caught it. She'd bought it at the store and used it for cleaning wax off the red and black geometric patterned canvas on her kitchen floor.

"It's too bad that the old man is dying," Aunt Callie said so matter-of-factly, she could have been saying, "The weather is *lunning*," or something as trivial.

Mandy pushed back her chair. If Issie died, he'd be lowered into the deep, dark ground full of worms and insects, then clobbered with wet, dark earth—heavy and cold, while his soul would get roasted in Hell. That's if what others said about him was any indication of what God thought of him. He hadn't been to church since he was carried there when he was a baby.

"Mandy!" Her aunt's voice was sharp, "Eat your liver."

"But I already have a liver," she answered, running from the kitchen into the hall and out through the door.

Mandy used to think that no one had ever died in her family. There were black and white pictures of her Gran and Pop Maley together with children big and small. Some had grown old and had quietly gone off to Heaven. She visualized them climbing Jacob's Ladder, then being helped by St. Peter onto a cloud. Now that she was older, she knew the difference. Lots of people in her family had died. Some were skeletons in the ground at the back of her grandmother's house. That's why there wasn't a potato garden there. There were crusted headstones toppling above some of the graves. She would die, too, if she wasn't caught up in The Rapture. She would be part of a picture like the one in church where people in flowing white gowns and bare feet, with raised hands, swim up from cemeteries into the air towards a white clad figure glowing with a yellow light around his head.

Mandy skirted around the back of the house so her aunt could not see her. She must save Issie. Sure he was old, but he didn't have to die yet. And his hands were steady. Issie could shave without a mirror; not even her father could do that. "Once you've gone over these grounds as often as I have," he'd chuckled, "you don't need a mirror."

Issie wouldn't wear a poppy on Remembrance Day, "'cause," he said sternly, to anyone trying to sell him one, "rich countries sell weapons to poor countries in order to make rich people richer, and then stir up common people to fight each other." He grumbled that a man was spotted if he didn't go to war and shoot someone, and he was a hero if he went and got shot himself. He might even get a Purple Heart. Issie said he had one heart and he didn't want to risk it for a purple one that was only good for making people feel proud of themselves for saving one lot of men while killing another lot.

Mandy knew there was a strange silence in the men who had been to war. It wasn't the inky tattoos staining their arms that marked them. There was a silence in them that seemed louder than a roar, and when they and Uncle Gus sat on the porch, they looked as if they were blind to the sun. Issie said that they had marched to the glory of war, and got blood on their hands. Then they came home and it seemed as if blood was still there by the way they kept looking at their hands.

Mandy ran as fast as she could across the lane that separated the two houses. A sob rose up from her insides and forced her mouth open in a sudden catch of breath. She ran over the matchboard door that still served as a bridge across the dark mouth of a brook. There were once boards there with spaces that she had fallen through, *rinding* her leg. "That's it, Girl," Issie had chuckled, "when you're looking up, you take the chance of falling down."

Sometimes she pretended he was the grandfather she never had. He never hugged her or said nice things to her, but when he looked at her she could see a gentleness in his face she knew must have been in her grandfather's. His hands were big and strong-looking. One day he was coming down the road with his hands apart. He had measured a window for a pane of glass and was on his way to the store. "Don't spake to I!" he cautioned, "you'll get me size all mixed up." Another time he'd gone to the store and asked for a box of wind for his daughter-in-law's wash day. "Not wind, " he'd added, "Breeze, that's the name of the suds."

If Aunt Callie's hens weren't laying, he'd send over some of his eggs. When he discovered that one of the hens was pecking the eggs and sucking them out, he burned its bill. Her aunt told Mandy that a whistling hen forebodes death and as she neared the house she imagined a hen crowing.

Issie had a cure for everything. He would burst the bladder on a tree and swallow the turpentine, and when he stuck an axe in his foot he filled the cut with *frankum* to stop the blood and keep out infection. Mandy thought he was both wonderful and disgusting when his granddaughter, Melanie, was choking. He'd never given much heed to the baby before, not even a cooch-e-choo, until the baby started to turn blue. Suddenly Issie grabbed the baby and started sucking on her nose. Then he spat in the coal bucket by the stove. "All stogged up, she was, with the cold," Issie explained. It seemed to be true. The baby turned pink almost immediately.

Mandy's freshly starched dress seemed to lose its freshness as she came close to the house. The smell of grease hit her nose.

When she pushed open the storm door and went into the porch she could see Bertha leaning against the back of a chair that had two rungs missing. If she doesn't give heed to her weight on the back of the chair, she might soon be sprawled on the canvas floor Mandy thought, noticing its red flowers scuffed to a dirty grey. Bertha's eyes were puffy and red. They usually were. She seemed to be wearing the same black nylon dress over a black slip that she had started wearing after her husband died two years before. Perhaps she has other dresses, Mandy thought benevolently. She just doesn't want to get them dirty like the black one. It showed crusty stains. When Bertha opened her mouth, it was as if she was unhinging long yellow nails with black flecks. The woman, though friendly enough, made Mandy want to hold her nose, and run out among her aunt's pine trees, gather some lilacs and shove them in Bertha's face. Instead she went in and sat on the lumpy daybed, pushed in a corner tight against the wall.

"Visiting your aunt, are you?" Bertha asked.

"Yes," was all Mandy could answer, her thoughts leaping with urgency towards the man upstairs, afraid he would die before she got a chance to tell him about Heaven.

A flash of silver caught her eye as she looked towards the window. "It's a wonder airplanes don't drop and squash us all," said Bertha, glancing towards a neighbour coming in the door.

Ignoring her remark, the neighbour said bluntly, "I suppose Issie will leave you some money."

"Leave it, Mary," Bertha's voice was abrupt, "he spent it—buying things for the kids. We'll miss that. He was getting a bit of old age pension, thanks be to God for Joey getting us into Confederation."

"He's nigh on going, is he?"

"Who knows? They say cracked pots last longest. The healthy ones up and goes, while ones you expect to clear out linger on. The old man, though, don't have much to hang around here for. He's been coughing for a month now as if he's got twine tangled up in his lungs. He's pretty quiet today though. His wind 'ill soon stop, I dare say."

"Maid, some old people can be a nuisance. Tarments, that's what they are. They goes right back to their childhood."

"Thank God, he didn't do that. His mind is as clear as spring water. It'll be some good cleaning out that room though. He's been there a long time and there's some dense cobwebs gathered in the bed frame. I'm just glad he wasn't laid up long enough to get bed sores. They're something shockin'. And then you got to rub *gentian violet* on them. I wouldn't have the stomach to face the likes of that. I'd have to let it bide."

She pulled down her jaw and scratched it with yellow fingernails; then she yawned like a foghorn. "I may throw out his mattress and move into his room yet. If he'd been like anyone else he would have gotten rid of the sack of shavings and feather mattress lying on ropes and always sagging like a hammock."

Mandy had seen the old man swing his grandchildren on his shoulders and trot around as if he was a horse. He gave them money for chips and drink. They won't miss Issie, thought Mandy. They'll go around as if he was never there. Perhaps they'll pick up a jacket or a stick of baccy and fling them in the stove to destroy the last smell of him, and if they miss him it will probably be with gladness at the difference in not having him around. You'd think they'd know that they'll be there some day, kicking against death while their relatives would like to shove them into its arms.

She loved him. Her feelings rose to a fullness she could hardly take, and she couldn't bear the thought of him slipping farther away every moment that passed. There was an urgency inside her to see him, to touch him—to say goodbye.

"Did you hear the buzz in the wall?" Mandy asked haltingly.

"The buzz?" Bertha and her neighbour exclaimed in unison.

Mandy started to explain, but Bertha waved her aside. "Oh that old superstition."

Mary said she'd better get on home. Mandy got up behind her, pretending she was going out the back door. Instead she sneaked up the stairs.

If she got frightened to death, it would be her own fault. "A sickroom is no place for a child," her mother once said.

Mandy crept onto the landing, then eased her way along to the room at the end of the hall. She shouldn't think they didn't care. Perhaps Bertha knew what it was like to feel death inside her when her husband died, and now she was through with letting death ruin the little bit of life she had left. Perhaps it was having death around that hardened her, made her anxious to have life fill the house. Maybe that was why she wanted her coughing father-in-law to go and take death with him. When it came right down to it, living came first.

The smell of liniment hit her nose. She could hear the old man's seesaw breathing as she entered the room fearfully. A cold feeling wrapped itself around her. The second Mandy saw him she knew why death was so terrible, why people were in pain when it got inside those they loved. Issie's head was lying on his pillow; a grey blanket wrapped him right up to his chin. His face was sucking in air and blowing it out in short gasps, leaving his cheeks hollow. She wanted to touch his silvery hair and whisper, but she was scared that her voice would drive his pain deeper into his flesh. She wished his mother would come to him now and let him out of his pain.

Aunt Callie had told her that Issie was only four when his mother died. His father married again, and he got a stepbrother who got the easy end of things, and a stepmother who didn't want some dead woman's son. One night when there was a full moon shining through his window, he saw his mother. He felt the weight of the bed go down when she sat on it. Later, when he got barred down in the cellar by his stepmother, he said that his mother came and let him out. He described her in a way that people remembered her, and they knew that what he said must be true because the hatch on the cellar was too high for him to reach.

One part of Mandy was on the lookout for Bertha and the other part of her was bursting to say something that would make Issie's time less alone. But first she must let the light into the cold, dark room. She walked toward the window and opened the venetian blinds so that the warmth of the sun could touch Issie.

Her fingers trembled in their haste to touch his face. It was soft and warm. Perhaps they were wrong, and he wasn't really dying. He stirred and mumbled a woman's name as if his memory was taking him away from all this, back to a time he loved.

It seemed that Mandy had known him all her life. She'd watched him in the stable—a strong man smelling of hay and the horses he brushed down. They snorted; he chortled. Some days he didn't talk much—just went along up the side of the hill gathering the hay into tight mounds. He was there during her summer holidays to give her a ride if she asked for one.

She was sorry she had come to say goodbye to him because now she would forget things about him, strong things. She would always see him lying with the breath almost gone out of him. But then she knew that time is always changing people into someone else. Issie was a baby, then he disappeared into a boy, then he became a handsome dark-haired man. She imagined him being so many people before she met him; she knew him only as an old man. His voice always seemed strong, as if his life was in it. But now he wasn't talking. Still, she could imagine him being so many things—except dead.

At first it was as if there was a curtain on his eyes, but as she talked to him, his eyes opened and it seemed as if all his life was there trying to get out—to reach her.

She wanted him to forget his pain and think of life. Death was cold like this room, ugly like the grave beneath the grave digger's shovel. "Heaven," she whispered falteringly against his half-closed eyes. "It is the best place in the whole wide world." She stopped to correct herself—"the best place in the universe." She knew she was saying the right thing because the minister had said there was no better place anywhere, and even though she wasn't as old as Issie, she wanted to go there as soon as possible.

"You won't need to pray to God up there. He'll be around if you want anything. The streets are gold, and there are no potholes. You can ride horses for fun; they don't need to be fed."

The old man's eyelids twitched as if he was straining to lift them. His lips, once strong and fine, were now pale and drawn across his teeth. Suddenly she saw them moving away from each other in a smile. He had heard her!

His lips backed together. A little puff of breath pushed them slightly open, then they closed. His life could not have gone out in that one puff. Was that all life was: a little puff of air let out, a tiny puff that barely opened one's lips? Mandy wondered.

And then the realization dawned. If life had gone, then something else had come: dark, cold, black death that no light could penetrate. Fear filled her. She jumped up as she heard voices. Heavy footsteps were coming across the landing. Issie's son, Jack, must have been sent for and now he was hurrying into the room. Bertha came behind him. She put her hand to her throat. "How did that child get up here?" she asked, letting out a *blaighard*. Mandy slid past her without a word and ran down the stairs. She stood at the bottom with a feeling of relief. Bertha's voice floated down to her, "He's in a trance."

"A trance, me arse; he's as dead as a nit," came Jack's abrupt answer. Mandy thought of how Issie had managed to take the blue out of the baby's face by sucking it. No one could take the grey out of his face. His skin had barely been able to hold in the pulsing blue veins in his hands.

Outside the house, the ocean was beating itself into a fury. Gulls overhead screamed as if taunting it on. "Dead!" she cried, running and sobbing into the wind, not knowing what to do with all the sadness that was inside her. As long as Issie was alive, it was the same as having a grandfather. The wind's fingers raked the hot tears from her face. She ran up the hill until her wobbly legs landed her like jelly on the hard knob facing the ocean. The sky was cold and grey with shattered pieces of dirty cloud. Suddenly she noticed a cloud that feathered like a giant bird wing.

"They shall mount up with wings as eagles," her father once said, quoting his Bible. Not like the bald eagle that was shot and mounted up on the scaffold at the cove entrance that she'd missed school to see.

Just then, a gull lifted itself from the eiderdown of gulls perched on a cliff. It looked as big as an eagle as it rose high into the air. Mandy caught the glimpse of a man in his boat lift his gun; a shot rang out. She felt the coldness, as if death had again reached out to frighten her. The bullet never touched the bird. It lifted higher and higher, soaring like an eagle. As she sat back to watch, it was as if the bird was an eagle taking the soul of Issie beyond the clouds to the heavens she'd told him about. Nothing could hold him back—especially not death.

"*Widdershins?*" *Mandy screwed up her face.* "*What's that?*"

Her aunt explained, "*That means he turned back the hands of the clock to the time Josh got the fright, then he wound the clock contrariwise until the spring broke. That was the doctor's way of breaking the spell. 'Twas a lovely grandfather clock...*"

WIDDERSHINS

It was Christmas Eve morning and Mandy knew exactly what would happen the moment Aunt Callie, tailed by Uncle Gus, pushed open the porch door and called, "Cheerio!" She looked at everyone sitting around the large wooden table and said, "Elizabeth, what a feast for sore eyes those children are!"

Mandy glanced at her mother's face. It looked flushed as she rushed around putting food on the table. Aunt Callie interrupted her with a big squeeze, and smiled plaintively, "Don't you think, Elizabeth, that you should share at Christmas? Let me take one of the children for a few days." She looked towards Jeff expectantly, but Jeff, who was now fifteen and in no mood to be shared, turned away. Michael was ten, and he, too, had no interest in spending Christmas at Smith's Point. Taylor seven, and four-year-old Timmy were more than Aunt Callie could handle. Her eyes swivelled to Mandy.

"Well," said Elizabeth without much heart, "if Mandy wants to spend Christmas with you..." Her voice trailed off.

It wasn't what Mandy wanted. Christmas meant being at home among squeals and laughter, as stockings were emptied, and red and green tissue paper was torn off mysterious-looking presents. How could a visit to her aunt's compare with that? There would be no tree, just strangers straggling in for a glass of grog. She'd be expected to sit on the settee in a large kitchen and keep quiet while strangers scrutinized her and discussed whose nose and eyes she might have come by.

She wanted to say, "I won't go." The words were almost pushed to the surface by a sense of rebellion, but the Christmas spirit must have seeped inside her conscience. She found herself thinking that there'd be no Christmas unless people shared. I suppose, she thought reluctantly, I could be a borrowed gift—one that is returned once Christmas is over.

She went to pack, then stopped suddenly as the thought of Smith's Point sent a shudder of loneliness through her. It was as if the seas just out from her aunt's house—as dark as the inside of a wolf's mouth; the wind as lonely as a wolf's howl—were all around her. If only it was summertime, when the sunshine and winds were warm. Then she liked to stand on the upstairs landings, and look down over the cliffs of land rising high above the waters. She'd watch the sea beat itself upon the rocks in an explosion of white spray, then fall to be enveloped again by its dark self. On a clear day she could look across to the point of land jutting out into the sea and take comfort that it held her house, even if it was on the far side and out of sight. Mandy shook her head, as if to knock some sense into it. After all, she was twelve— just the age to be thinking about someone other than herself.

Mandy's mother came through the hall and stood in the doorway. "You be good while you're gone," she said, pushing her hands down over her middle as if to flatten her large stomach. A frown crossed her face. "And stay clear of old Josh."

She could have asked why, but her mother was apt to tell her to bide by rules without questions and answers. "Too much knowledge," she warned, "could clear out a child's innocence."

Mandy did know some things. She knew that Old Josh was Uncle Gus's brother, and he was nicknamed "loser boozer." As she was leaving her room she stopped. Her parents were in the hall. Her mother's voice carried an edge of worry. "I hope Mandy won't be afraid of the drunks Gus tows home. Old Josh will probably be around sniffing for a bellyful of liquor, while his woman and that little girl of his starves in the paper shack he's hauled over their heads."

"And after what happened to the boy—the man's a killer." Her father sighed heavily. Their voices turned to something else as Mandy came into the hall.

It was a half hour's ride to Aunt Callie's, along the shore, then up over hills that dipped into valleys. The final ten minutes were spent following a road that edged the high cliffs like a pencil mark. Mandy shuddered as she remembered that on a dark night last winter a man had gone over the cliffs and died.

The two-storey witch hat house was cold, and it seemed empty, perhaps, because there were only the voices the three of them brought into it. Once the black stove got some fire into its insides the air began to warm up. Soon the water in the kettle began to gallop and spout, making tissing sounds as beads of it popped along the damper. It began to seem more like home, and Mandy, feeling the heat seep into her bones, moved to read Christmas cards strung across the wall on red twine. She liked the ones puffed with powder under silk bells and ribbons, and those with glitter on them.

Aunt Callie called her away from the cards to sit for milky tea and bread with partridgeberry jam and tin cream. After they'd eaten, Mandy was left to wash the dishes while her aunt went into the pantry for her cap. She pushed her frizzed, grey hair up under its edge and began to mix some molasses buns. Now that she was past seventy, her cheeks were becoming more and more like shrivelled potato skin; but her eyes were bright, almost as if they were sitting in water.

"Are you making buns for company?" asked Mandy.

"Yes, Child, I expect quality every time I leave the cover off the tea pot." Aunt Callie called company quality and, at such times, served her best food on her good dishes, spread on her finest tablecloths, made from flour sacks and bleached to a dazzling whiteness. Each cloth was embroidered from corner to corner with flowers and fancy edging.

"Praise be!" Aunt Callie said suddenly. "There's the crowd of Gus's buddies coming for their swigs. That's the rigmarole on Christmas Eve."

Mandy skittered out of the kitchen and into the parlour just as the front doors opened. She stayed still in Aunt Callie's big chair hoping the roars of laughter wouldn't come any closer. Suddenly the door creaked and opened. A man who looked a lot younger than her uncle, came into the room. "I heard about you," he said boldly, "the girl with the turned-up nose and horse's mane of hair." Mandy pressed her back against the back of the chair as he came closer, his face flushed, and his eyes bright. There was a strange smell on his breath, like the hops her aunt used to help make her bread rise.

"Would you like some money?" he asked, his breath hot against her face. He dropped some coins into her lap. "I could spit on them for good luck," he grinned.

Mandy tried to answer indignantly, but her voice came in a hoarse whisper, "No, thank you."

He laughed, then pushed his hand down under the money in her lap. She stood up with a sudden movement, and the man tipped back on his heels. The money fell to the floor, some of it standing on end in the braided mat. There was a sound outside the door and the intruder hurried over to the mantle where he stood whistling, and pretending to look at her aunt's and uncle's wedding picture. Mandy's shoulders sagged in relief when her aunt came into the room and said in a stern voice, "Come along young man—out among your own crowd."

Mandy was glad when the door slammed for the last time and the house quietened down. It was like being able to breathe again. Then there came the sound of the porch door being unlatched. "That's your company now," said Mandy.

"I've had company," sniffed Aunt Callie, "and a fine lot they were—left the house, they did, loosened up like rag dolls. No, that's Bertha from next door. I was looking out the pantry window and I saw her coming down the lane."

"Nice clear afternoon for Christmas Eve," called Bertha, sticking her head in around the door facing. "Josh is drunk again. I saw him trailing all over the road. I heard he's beating Jane again, and only six months after poor little Joey died. 'Tis an awful sight to see her with black eyes against her white, drawn skin." Bertha sniffed and her lip curled up towards her nose. She pushed a heavy hand against it, then rubbed it down her bumpy hip. She didn't need to say, "I got a cold—I have."

"Then don't do what Gus did," warned Aunt Callie. "He sniffed *hyssop salts* up his nose, and emptied his head of everything—even his brains, I declare."

"I'm not that drastic," snorted Bertha. She pulled her bulky body over to the window. "There he goes—Old Josh—off to the house smelling for a grog. He won't know he has a soul when he's finished, and the little one and that woman of his won't know it's Christmas. I better get on over to the house and lock the door, just in case he shows up."

"A *beggar's claptrap*, that's what she is," Aunt Callie said, once Bertha's back was in the distance, "always minding other people's accounts."

"Salt herring," said Uncle Gus, with relish, as he lifted the cover off the steaming pot. He sniffed the steam. "Ha—'tis better than a rock in a pot for Christmas Eve." He turned and took hold of Mandy's hair, "And what are you expecting from Sandy Claus?"

"Nothing, Uncle," she answered, pulling her hair out of his hand.

"Nothing!" he raised his voice. "We won't have that here—that's a hole with no sides."

Aunt Callie's eyes took on a thoughtful look. She left off setting the table, and stood for awhile with her knuckles on her broad hips, making her arms stick out like handles on a jug. "I think," she said with a nod, "we'll dodge down to the shop."

"Will we see Alice?" asked Mandy eagerly.

"Alice?"

"Yes, Old Josh's little girl. Where does she live?"

"In the brown house under the hill," answered Uncle Gus. He lifted his eyebrows, then he said with a straight face "—in the one that's built outdoors."

"Knock off your foolishness," said her aunt, "and don't stretch your suspenders."

Uncle Gus's molasses-warm eyes turned muddy brown. "Hush woman, if I wanted to get hen-pecked I'd go down to the hen house." He pulled a block of Lady twist tobacco from his pocket and twisted a piece off with his teeth.

Mandy could have laughed at him chewing on his cud, and running his brown, hairy hand over the top of his head that, with its sparse strawy hair sticking off, looked like a coconut. But she'd rather die than risk having him squirt his baccy juice over her clean dress. She hurried to get her coat.

"Do you think we could buy some apples with points?" asked Mandy, as they trudged through the deep snow.

"Indeed we can; I've a few cents for that."

"When you cut them in half," Mandy said, her blue eyes widening, "you can see the Star of Bethlehem."

"That's the ones," her aunt replied, "and I've saved enough money to buy a duck for Christmas dinner. We'll have Alice and Jane up, and Josh, if he can get one foot in front of the other. In some ways, that man is like his mother, God rest her soul—if he can. She was a bad one. Living with her in this house was like having a foot in Hell, which is what she must have had when she was dying. She sat in her rocking chair and asked for water. Before I could get the glass to her lips she was gone. It was wonderful not having to stay out of her way for fear a pot of scalding water would be thrown at me. That's not even the half—but never mind. 'Tis Christmas and now I'm being a claptrap."

While her aunt was picking out some apples and a large duck, Mandy bought some hard knob sweets from the big round bottle on the counter. She turned quickly when she heard her aunt say, "Hello Alice, how's your mother?" Now she remembered her. She'd seen her once when she and Aunt Callie had gone to Long Cove. That's where the family lived before the house burned. Alice's thin, yellow hair hung in strings around her pale face, and across the collar of a coat that looked as if it had been turned to hide the faded colour.

"She's okay, Aunt," Alice answered timidly.

"Only a few hours 'til Christmas," Mandy smiled.

"I can count forever," Alice answered sadly, her eyes like blue buttons in ragged, black-stitched button holes.

"But it's tomorrow," said Mandy, excitement rising in her voice.

Alice shook her bent head; tangled hair falling into her eyes, made her look like a little rag-moll.

"It will come, if you believe," Mandy tried to tell her.

Alice lifted her head, and a flicker of hope came into her eyes, then died.

"Come along, Mandy," said Aunt Callie briskly. She turned back to Alice. "Tell your mother we're expecting all of you for dinner."

"Yes, Aunt," Alice's voice quavered as if she was uncertain.

As they walked home, the falling snow enclosed Mandy and her aunt like the tossed flakes in a water globe. Just below Aunt Callie's house Mandy noticed a dark patch on the snow. The evening was darkening, but her aunt knew what it was right away.

"That," said Aunt Callie in an indignant voice, "is Josh. He's done it again. Not only does he not keep Christmas; he does his best to try to steal it from everyone else. I'll ask Gus to help him get to the house; otherwise, he'll perish. We'll stretch him on the daybed and let him sleep off the only Christmas spirits he knows anything about. If he comes to, and acts up, Gus will put a fist to his jaw."

Mandy had never seen anyone drunk before. The thought of being in a house with a drunk man made her as skittish as a frightened horse, and she started to run past the curled-up figure, past her aunt's house, and across the lane to Bertha's. She paid no heed to her aunt's call to wait. The sound of running brought Bertha to the door, and Mandy fell into her arms panting, "There's a man—Old Josh—he's coming to Aunt Callie's house."

"You're right in feeling scared," said Bertha, drawing her inside. "Drunks," she continued, with a gleam in her eye, "are the only spirits of Christmas—take my word. I've had strangers that I've never laid eyes on before, pounding the door. Like mummers, they were, huffing to get in."

Mandy and Bertha watched from behind the curtains as Gus dragged his brother towards the house. Mandy's voice quavered, "I can't go out there—he'll grab me."

"He's done some terrible things," said Bertha. "I heard your uncle say, 'Josh, you got it coming to you'. Retribution, it's called; it's on the way. Billy, the oldest, moved out last year. Every now and then he fires a rock through the window, hoping to hit the old man."

"That's awful!" Mandy said, cringing.

"It seems that way, but Billy learned evil from his father. He pushed him down the stairs one night, and told him he hoped he'd never walk again. Other times, he'd come home late at night and dribble over the children's beds. Then he'd push them out of bed, and order them out the door, sometimes on the wildest winter nights. After that, he'd crawl into bed beside Jane, and sleep like a baby. If he has a conscience, no one has ever seen a sign of it—unless 'tis the devil." Bertha pulled down her face making her double chin triple against her fat neck. "When he's drunk, 'tis as if the devil comes to life in him as plain as day."

Mandy stared at Bertha, her eyes brimming with tears, "Why didn't Alice's mother stop him?"

"She tried—but she was no match for Josh's fists. One time, he knocked her out with a *trotter bone*. She wants to stay alive for the children. Your aunt always took Joey in, and Alice too, when she didn't find some place to hide. I minds one time when your aunt and uncle were away; Joey came knocking on my door. He was a small frame in his one piece underwear, and he stood on my step in his bare feet jumping from foot to foot and shivering like a little lamb. I took him in for the night. Then last January he took with meningitis and—only God knows what else. By the time he died in June he was curled up something terrible; his little hands were bent like lobster tails. He was thirteen—an unlucky number for him. Old Josh sat at the wake with his elbows on his knees, and the backs of his hands under his chin. I couldn't help speaking my mind. I asked him if he remembered all the nights he threw Joey out the door. Then I said, 'Now he's outdoors for good.' I had to get up and leave for fear he'd strike me; but like any decent soul would, I got my word in. Since then, Alice doesn't get disturbed. Perhaps he mellowed."

Mandy jumped at the sound of a knock at the door. The knob turned, the door opened and her Uncle Gus came in. "Come on," he said briskly, "home with you." He grabbed her hand and held it tight. Mandy went without a word. Once she was inside the porch, she dropped back behind her uncle. Fear tightened around her spine like a fist when she looked across to the daybed and saw a man stretched out, and snoring loudly—his mouth open like a black hole. His face, under a head of hair like black sheep wool, looked yellow and cracked, as if an earthquake had shaken it.

"Hurry, Child," said Aunt Callie, "up to bed with you, so you can get an early start on Christmas day."

"I can't—Aunt—I can't pass him," her voice faltered.

Aunt Callie held out her hand, and Mandy took it. She held herself tight against her aunt's body, away from the man on the daybed. Just as they passed him, he let out a snore that almost dropped Mandy to her knees. Fear bit into her backbone and she ran into the hall. She caught hold of the stair railing, hitching her toes in the metal clips that nailed the stair canvas.

"That man can't hurt you," said Aunt Callie.

"But Bertha said —"

"Bertha!" her aunt shook her head. "I already told you she's a tongue wagger."

"Are they true, Aunt Callie—all those terrible things?" asked Mandy, once she was on the stair landing.

"'Tis true, Lawd Gawd," her uncle muttered heavily, as he came up the stairs. "If I let myself think about it, I'd heave his carcass out the door, and let him bide to freeze."

"Stop trembling, Child," said Aunt Callie, "no one should be afraid on Christmas Eve. She drew Mandy against her warm, hot-water-bottle softness, then she took her hand and led her over to sit on the wooden trunk under the window. "Let me tell you about Josh," she said gently. "He was young once—about your age when it happened."

"It happened?" Mandy looked at her aunt and frowned.

"Josh had stayed too long at his cousin's house. It was getting dark, and he decided to take a short cut across the marsh. According to him, an old woman with an Aladdin's lamp caught hold of his hand. She didn't let go until they had crossed the marsh, then she disappeared. When Josh got home he was in a state of shock. The fright must have turned his blood because he swelled up like a poisoned rat. His mother had to press down his tongue to get in water. When the old doctor came, he told Gus— he was only a youngster then—to run quick and get some twigs. The doctor twisted the twigs into a cross and then made the sign of the cross over Josh. He asked what time the spell had taken him. Then he shook the clock and did *widdershins.*"

"Widdershins?" Mandy screwed up her face. "What's that?"

Her aunt explained, "That means he turned back the hands of the clock to the time Josh got the fright, then he wound the clock contrariwise until the spring broke. That was the doctor's way of breaking the spell. 'Twas a lovely grandfather clock, but it never told the time again. It was kicked around outdoors for a long time. The doctor left a flask of whisky for Josh's nerve. He told his mother to keep the flask in his pocket every day, so his nerve would get strong. The doctor didn't say when she could stop, and she didn't ask. Months went by, then years, and Josh was still nipping at the bottle." Aunt Callie's eyes got angry-looking when she said, "He got frightened by one spirit, then ended up with another one fixed to him for life."

"Whose fault is it?" asked Mandy.

"The world isn't perfect, Child," her aunt sighed. "In a way Josh is only as bad as he was made. We have to do our bit to help everyone who suffers. That's why your uncle, for all his barking about Josh, does what he can to help out." Aunt Callie raised an eyebrow in Uncle Gus's direction. He shrugged, and reminded her that there was a surprise for Mandy down in the parlour.

Mandy followed her aunt back downstairs, with Uncle Gus coming behind. She peeped cautiously into the room. "A Christmas tree," she whispered, then her mouth dropped open, "it's bare."

"Just as God made it," said Aunt Callie with a firm mouth, "not fancied up—or false."

Uncle Gus snorted, "Now Maid, don't confuse the child. God made you bare, but you wouldn't be caught in the open that way. A few cranberries on some Christmas twine might have brightened it up a bit."

Aunt Callie ignored his suggestion. Her eyes flashed over his remarks about her. "Your mind, Gus, 'tis always in the devil's workshop."

Across the hall, Josh let out a snore. Mandy turned in fright, and started back up the stairs. Her aunt called after her, "Go to bed, my dear. These are wonderful times. When I was a girl, things were bad. I was a servant girl at your age—up at four in the morning—drawing water from the well, two buckets at a time. I fell asleep more than once, with my hands in the bread dough." She flapped her hand as if to shoo Mandy off to bed, "Never mind; that's all behind."

Mandy burrowed deep into the feathered bed, letting her feet rest on the round, smooth rock that her aunt had warmed and wrapped in flannelette. Her thoughts went out like a light, and she slept.

She awoke suddenly, something touching her senses; excitement came like a ripple widening inside of her. She sat up quickly, pushing the fat feather pillow behind her back, then she reached towards the stuffed wool *vamp* lying across the foot of her bed. The first thing her hand touched was an apple. She rubbed it against her nightie to make it shine, then she bit off one point, drawing the sweet Christmassy scent up her nose. Then she laid it aside, and reached to take out an orange wrapped in its own tissue paper. After that she drew out a present wrapped in red tissue paper and tied with green twine. She tipped the vamp and let some sweets fall out of its toe. She closed her eyes so she could better hear the mysterious rustle of the paper as she tore it off. A pretty cotton apron with a frilly bib fell out. There were pictures of kittens all over it. A thoughtful look came into her eyes. "Alice would love this Christmas box, and anyone can be Santa Claus—a man, a woman—even me!" she exclaimed. Mandy jumped out of bed and pulled on her brown ribbed stockings. She shivered her way into her red wool dress, then stood on the hooked mat as she did her best to rewrap the present. She put the apple and some sweets into her pocket. The window was a silver plate of designs in wings, and horses' manes, and even a silver tree, through which she blew a black hole, so she could see outside. It must have rained rhinestones, she thought, looking down at the snow glittering in the sunlight. She could see conkerbells, or ice candles, as her aunt called them, hanging from the eaves of Bertha's house. Tree branches held up fingers encased in silver gloves.

She turned from the window and tiptoed out to the landing. Silence was all around—like a breath kept in. She tried not to think of Josh as she slid her hands carefully down the stair railing, keeping the weight off her feet so the stairs wouldn't creak. She turned her face away as she pushed open the door and passed Josh. He let out a loud snore and she jumped in fright. She hurried into the pantry. There she took a knife and cut her apple in half, hoping that Alice would see the star of Bethlehem in the centre. The half-apple was wrapped in a flour bag cloth and pushed back into her pocket. She couldn't look at Josh as she went into the hall to get her coat.

Mandy walked across the crunchy snow in her white, fur top boots. Blossoms of snow began to fall thick and fast covering her hair like a lace scarf. It was the most beautiful Christmas morning world she had ever walked into.

This must be the house Alice is living in now, thought Mandy, coming to an old brown bungalow. She reached up and lifted the latch on the grey matchbox door. She pushed it open, holding her breath for fear its hinges would squeak. There were no sounds, only a cold emptiness as she went through the porch into the big kitchen. Alice lay asleep on the daybed under old coats and a quilt like the one on the bed Mandy was sleeping in. Her hand was under her chin. Mandy placed the present, and the apple and sweets on the bed, noticing that the contents in the chamber pot by the bed were frozen solid.

Mandy hurried back to her aunt's house. This time she forced herself to stop and look at Josh. He looked, as her aunt would say, like a poor mortal, rather than an evil man. To prove she didn't have to be afraid, she reached out a trembling hand and gingerly touched a nose that was sprouting hairs. Probably fertilized by the booze, she thought. His eyelids lifted slightly, then dropped again. "Merry Christmas," she said hoarsely. Then she turned and crept upstairs and into bed to wait for the stove fire to be lit.

"Do you like your apron?" her aunt asked, when she came downstairs for breakfast.

"Yes," answered Mandy, "it was beautiful."

"Was—where is it?"

"I gave it away."

"You gave it away?"

"Yes, Aunt—to Alice. I went over this morning."

"Oh," her eyebrows lifted, and her eyes looked thoughtful as she turned to crack two brown shell eggs into the frying pan. "Why?"

"Because, well," she went on in a rush, "my dress doesn't get dirty if I'm careful. I don't really need it."

"And Alice needs it?" her aunt asked gently.

"I don't know, Aunt Callie." She moved in front of her, and said with conviction, "She needs Christmas."

Her aunt smiled and said quietly, "Since Alice already has her Christmas box, the one under the tree must be yours."

Mandy went into the parlour and lifted the present from under the tree. She tore off the green twine and opened the red tissue paper. She smoothed out her apron so that she could see the kittens that looked as if they were playing in her lap. I guess, thought Mandy, sitting in front of the tree, Christmas away from home isn't so bad, not when you give away your present and get back one exactly like it.

... daydreaming could take her away from now and back to times when the sea brought a romantic edge to the lives of people living near it. Tales were written of women at windows, curtains hanging down their faces like white mantillas. Their eyes burned with hope and fear as they speared the distance...

THE LADY IN BLUE

The ocean fitted into the frame of Mandy's classroom window like a painting that magically changed moods as often as its artist. Some days she was blind to its waters rising into the air in sudden splashes: white-tipped waves falling in foam and separating, as if melting into the cold blue sea. Other days, her gaze was drawn by sparkling flecks of sunshine playing on the gently rippling waters, its dust lightening the waters to a cool green. Often, the put-a-putt sounds of skiffs, rocking gently across the bay like toy boats, pulled her ears away from the teacher's voice. Her eyes lifted dreamily towards the delicate line between the sea and sky as if they were two globes waiting to discover sights and sounds beyond the place that seemed to hold her in a glass paperweight existence. Sometimes when the classroom window was open she felt propelled by warm, sweet-smelling winds. Today, she imagined herself drifting overhead with the seagulls. Suddenly she was dropped back into her seat by the sharp sting of the teacher's strap on her fingers.

It was hard to stay grounded when daydreaming could take her away from now and back to times when the sea brought a romantic edge to the lives of people living near it. Tales were written of women at windows, curtains hanging down their faces like white mantillas. Their eyes burned with hope and fear as they speared the distance for some sign that the men who went fishing that morning would return.

There was a painting hanging in the hall of Mandy's house that had become a part of her. Its heavy frame, overlaid with tarnished gold scrolling, was the perfect setting for the slender woman standing on a beach wearing a long, blue dress. A fisherman's net was draped around her shoulders like a shawl. She'd looked at the painting long enough to actually see the sea move back and forth, and come towards the woman's feet, where it beat itself into the white foam that bordered the beach like crocheted lace. The hand lifted to her forehead was delicate white. It curved over her eyes as if to carry her gaze beyond the black boiling sea. Though Mandy couldn't see her face, she imagined it to be young and beautiful, with blue eyes heavy with love, as the fisherman's wife waited and longed for her handsome husband to come home.

It wasn't like that in Maley's Cove. Mandy's mother, Elizabeth Maley, wasn't a slender lady in a beautiful dress, waiting for her husband to come home. She was often down in the cove helping Mandy's father, Eric, and Uncle George split, gut and salt the codfish that *The Flyer* brought in. She may have already been down in the cove today to wash out salt-bulked fish. Mandy pictured her mother's short, corseted body stuffed into a pair of men's trousers, which were wrapped tightly around her legs; her legs were then stuffed into a pair of black rubbers. Shabby orlon sweaters, one inside the other, strained against her large breasts. A nylon scarf was wrapped around her head and tied under her hair which was rolled on a piece of electric wire and shaped against the nape of her neck like a sausage.

Through her bedroom window, Mandy watched many times as her mother moved heavily down the road, past the bend, and out of sight, on her way to the fish flakes that edged the cliff like stone-face gods that were spit at or rocked against, depending on the sea's mood. She was glad she didn't live on the edge of cliffs, as did many of the cove people, where the lonely sounds of the sea, and the whistling winds turned them in their sleep with strange dreams. Her home was nestled in a hollow of barren hills, sloping away from dug-out cliffs, and sheltered from the bite of the easterly wind.

As the final days of June drew near, Mandy turned away from the school window framing the sea, away from her daydreams about the lady in blue and burst out of school into the shimmering heat of the summer sun. She felt the same sense of freedom that came with spring when she exchanged her winter boots for shoes and skipped over the hard-caked roads. Only she wasn't free. Timmy, four, was the youngest of her four brothers. He followed Mandy around like a dog's tail. She had to take care of him during the summer months while her mother spent more time at the fish.

Today she was awakened while it was still dark, by the murmur of voices and the sounds of her father's heavy feet moving through the hall. He and Uncle George had taken in their trawls. Now they had to set cod traps in berths in Cape St. Cabe. They wouldn't be *bulking* fish in the stage. They'd be emptying the hold of *The Flyer* on the big wharf in Bayview, where they'd split and gut the fish to sell to the plant there.

Mandy's mother liked to start housework just as the orange ball of the sun rose above the water, so she'd be ready for "the call." By the time Old Ned, the cove's taximan, blew his horn outside the door, she'd have her housework done, and was ready to pull on her fishing clothes. Sometimes Mandy would go along, taking Timmy. Fifteen-year-old Jeff was already in the boat, while Michael and Taylor stayed around the cove, not far from Grandmother Maley's watchful eye. And along the way, they'd pick up Aunt Caroline.

Today Aunt Caroline wasn't quite ready when Old Ned came for her, and by the time they got to Bayview, Mandy's father and Uncle George had already begun to split codfish. With bulging eyes in an angular head, the codfish looked as if it had been frightened to death, rather than lured into a codtrap. It was strange how its long, loose belly reminded Mandy of her mother's white calves. Mandy was glad her mother couldn't read her thoughts, or sense the discomforting feelings she carried about her mother's life.

Her mother pulled the bib of her new *barbel* over her head. Made from bleached flour sacks, and soaked in linseed oil, it had also been dipped in bright blue paint to give it a brighter color than the usual putrid yellow other fishermen's wives settled for. Both her mother and Aunt Caroline were anxious to start on the 60 *cantals* of fish brought in by *The Flyer*. They were soon standing beside their men and pulling the insides of the fish loose and letting them drop at their feet. Her mother rarely spoke as she stood in her black rubbers, ankle-deep in a gelatinous mass of fish entrails, her barbel splattered like the artist's palette Jeff had at home.

As Elizabeth pushed her gloved hand against the inside of the fish and pulled its entrails towards her, a gall bladder burst. Its green fluid shot her in the mouth. Mandy had left Timmy with Jeff and she was pronging fish into the box when she noticed her mother's face turning grey. She watched in embarrassment as her breasts heaved and her mouth opened. Suddenly her mother's lunch was flushed up and out among the slub at her feet.

The women at the next table, who had been laughing and talking, looked towards Mandy's mother, then turned back to their gutting. As her mother went to get some water, Mandy heard one of the women say, "There's going to be a few babies this year."

"Yes," nodded another. "There's going to be six when they all turn up—and perhaps another." The women looked towards Elizabeth who was wiping her face.

Her mother—she couldn't—! Aunt Caroline's voice drew Mandy away from the shock. "Here," she called, "skin up your sleeves, and gut some fish while your mother gets her bearings."

Mandy's mother came back to the table and, without a word, pulled on her goried gloves and barbel. Almost suddenly, the sun dropped below the rim of the sea, stripping the air of all warmth. Mandy saw her father glance at her mother's face, looking bruised from the cold wind. He stopped working long enough to rig up an old sail to shelter them from the wind that was beginning to moisten with rain.

Finally the fish were all done, and Mandy's father started hosing down the tables. By now the damp air was packed against her skin, sending shivers through every pore. She was relieved when *The Flyer* finally pulled away from the wharf, glad it was some other boat that would be there until midnight. Lights from the plant followed them with a single shaft of light. Her father and Uncle George soon let their heavy weight fall into the bunks in the cuddy while Jeff steered the boat towards home. Mandy's mother and Aunt Caroline set about getting a mug-up for the men.

Mandy leaned against the sloping back of the cuddy and closed her eyes. The lady in blue probably had two children: a boy and girl—and a husband who pulled her into his arms with a glad look. She never drew back like Mandy's mother did when her father sat down and pulled her towards his knee. Mandy noticed that whenever her father wrapped his strong, hairy arms around her mother, she pushed them loose and stood up murmuring, "Oh Eric!" A flushed, uneasy look came over her face.

Mandy was puzzled. She wondered how two people could spend a day hardly speaking—at least, not in endearments—and then in the darkness touch each other and be together. A picture of her mother's stomach flashed before her eyes. It did seem fuller. Her body was always rounded by flesh-toned corsets and laced. Corset stays ran up through them, bent with her mother's weight enough to look like barrel staves. This gave Elizabeth a barrelled look from the waist down. Her breasts hung over her corsets like dough left too long in the bread pan. Mandy could imagine the laces being let loose slowly—as the months passed—until the corsets were finally discarded, and large tops would have to be worn over shapeless skirts. That was the way it was when her mother was having Timmy. She moved clumsily for a long time.

Now she knew why her mother was keeping the baby clothes. She could almost smell the white substance she called starch burning in a pan on the stove until it turned dirty yellow. Mandy remembered her sprinkling it on the belly band that was pinned around Timmy until he was rid of the ugly blob that hid his real belly button. When Mandy was little, her mother used to say that she shouldn't pick at her belly button—and she wouldn't, for fear that it would come undone and her belly keg would open. Later, as she learned more, she figured that the belly button was there for unwinding as a baby grew inside the belly. Finally the belly would open like the mouth of a pudding bag to let the baby out, then close up again before anything else fell out. She no longer thought that way, now that she'd seen the doctor's book.

"It's gone," she'd said flatly, when her mother had asked if she'd seen the box of baby clothes that was in the closet. "We don't need any more babies in this house."

"That's something you don't know." Her mother's voice was mild and her face looked relaxed as if she really wanted another baby. "There's always room for love."

"Room for more babysitting," Mandy retorted. "I want new shoes, and a new dress, not a baby." Her anger swelled inside her and her voice rose. "I knew you were keeping the clothes for something—I'll burn them—I will." Her voice rose higher and higher. When her mother left the room, she pulled the box from under her bed, and gathered the soft folds of clothes towards her. She dropped them back into the box admitting aloud, "I'm almost thirteen. I know that burning the clothes won't stop the baby."

The day her mother fell was the day Mandy decided to accept a baby sister—or brother. It was a sunny day with tourists sniffing around the wharf and screwing up their faces at the sight of the *britchins* Mandy was picking out of the fish entrails. One woman in a white dress and high heels looked at Mandy and said disdainfully, "Cod roe looks absolutely vulgar. Anyone eating something that resembles pink, long-legged bloomers blowing, full of wind, on a clothes line, has depraved taste buds."

Mandy heard her name called sharply. She knew immediately what her mother wanted. Mandy knew that her mother never expected to be a tourist, or whatever else people who took holidays, were called. She looked uncomfortable standing on the wharf in a goried barbel, with splatters of fish blood on her face, while well-dressed tourists watched. She wanted Mandy to stop talking to them so they would move on.

The fish had all been gutted, and her mother was pushing the last wheelbarrow of cod to the weigh-in. The wheels cut ribbons through the slime that greased the wharf. Suddenly Elizabeth's arms flew into the air. She fell back on the concrete floor beside the weighing scales.

A shock of pain ripped through Mandy, forcing out a sharp cry that blended with her mother's cry before she lay still. Mandy watched not moving, her fingernails bit into her palms like shards of glass; her eyes and nose stung with tears. She should have kept quiet about the baby. Now it was bumped—or ...? A twinge of concern was pushed back as a feeling of rebellion swelled in her. She just wouldn't think about it. Then her conscience would have to be quiet.

Her mother was carefully lifted down into the cuddy of the boat. She lay there shivering while Mandy's father hurried to light the bogey stove. He reached into the shelf for a knife, and cut the apron strings from her mother's barbel. He balled up the barbel and shoved it under the seat. Mandy watched him sit down and take one of her mother's hands in his. He rubbed her small, limp hand between his big hands. Suddenly he reached to touch her face. "Like apples, your cheeks were, when we got married," he said, in a voice so soft, Mandy could only stare at him.

Unable to bear her mother's silence, Mandy grabbed the barbel, quickly climbed the *strouters* to the wharf and threw it over. She watched it drift out into the bay. Suddenly she was sobbing against Aunt Caroline's shoulder. "Mom won't ever wear this barbel again."

Her aunt lifted her chin and spoke firmly, "Those bright, young eyes of yours are straining to see the world and its glories beyond this island, but this is your mother's life, her chance to be away from the house where her work never brings her money. There's nothing glamorous about this job, but the money is clean—and accidents can happen anywhere."

Uncle George looked towards them. "You ought not to bawl this way." His voice was gruff, "How old are you—twelve—thirteen?"

Mandy turned her wet face to look at him. "I'm thirteen today."

"Did anyone wish you Happy Birthday?" Aunt Caroline asked kindly.

"Mom forgot," she mumbled, wiping her face with her fingers. She turned to watch a doctor hurrying aboard the boat.

Her ears strained to hear what the doctor was saying to Aunt Caroline, after he came up from examining her mother, but his voice was too low.

As the boat steamed towards home, Mandy sat with Timmy's head in her lap. Aunt Caroline's quietness frightened her. She took a piece of wet flour bag and wiped the freckles of fish blood from her mother's face. "I'm sorry, Mom," she whispered. "The baby's clothes are under the floor boards in the closet. I was going to tell you as soon as the baby came." She kissed her mother's cheek. It was as cold as a cod's nose.

"She'll be all right," Mandy said stubbornly, searching her father's clouded face for signs of hope. Instead she heard her mother's muffled voice, "I'm resting—just resting."

Mandy swallowed her tears and said gladly, "I know, Mom, I know."

Weeks seemed to hold on forever. Her mother lay in bed too sick to be moved over the washboard roads to a hospital. Sometimes she mumbled about roaring seas, and billowing waves lifting her high, then slamming her down. Dr. Evans blamed her feelings of seasickness on an ear infection, but the medicine he left changed nothing.

Mandy coaxed her mother to take soup and juices, but they didn't help her get better. She lay there as if immersed in a deep fog. Her eyes were shadowed and her lids drooped. Her mouth dropped open like the mouth of Timmy's sock, once the elastic had let go.

"Is the baby almost made?" Mandy asked Aunt Caroline, who just smiled and nodded. It had always been easy for her to talk to her aunt. She didn't mind telling her things that she didn't even ask about. Aunt Caroline said that men are restless and strong-natured during the winter months. That was why there were so many fishermen's wives with babies growing inside them during the fishing season. When Mandy queried further, her aunt had laughed. It was Aunt Caroline who had told her about the flour sack strips tied to bed posts so that a woman about to have a baby could hold on to them when the pain came. She had laughed at Mandy's shocked expression. "Your Aunt Caroline," her mother once warned, "always speaks whatever comes into her head."

Her mother was different. Mandy hardly ever knew what she was thinking and feeling, even while she held a crying baby against her heart. Her mother was the short sturdy tree they'd all leaned against. Now the tree had fallen, leaving a wide open space through which a cold wind blew.

The closet between Mandy's room and her parents' bedroom was a hall with curtains at both entrances. Fearing some strange presence in the house, Mandy couldn't resist the temptation to press her ear to the *ten-test* walls and listen to her parents' murmuring voices. One night her mother's voice came strong enough for her to hear her say, "It was so dark. Then suddenly there was a headstone in front of me—all lit up. I strained to see the name." She stopped, as if to catch her breath. "It was my name; I wasn't afraid."

"Elizabeth!" Her father's voice was sharp, "don't say that; you're not going to die."

Die! The word hit Mandy like a sliver of ice cutting through her veins. She skittered back to bed as if death itself was stalking her. She now knew what the presence was that had penetrated the house, moving in it like a ghost.

During those weeks, she stopped looking at the lady in blue when she passed through the hall. What did romance matter now? She often touched her mother's forehead. It was hot and her cheeks looked as if they'd been near a hot burner. She wasn't eating now. All she wanted was juice.

It was a cold, dark evening. Rain lashed at the windows, and moaning sounds swept through the house. Mandy and her brothers were sitting at the table eating supper in silence. Their father came through the hall, closed the door and stood in front of them. His heavy breathing seemed to sweep over the table and settle heavily. Even Timmy looked up as if sensing danger.

"Your mother," he choked, as if holding a mouthful of water. "She's—I don't believe she's going to make it." He moved around the table as if blinded. His faded blue eyes, usually wind-dry, were now bright and red-rimmed in his heavy, stubbed face as he kissed each of the boys. When he came to Mandy, she pushed back her chair and stood up, pulling away from him, as if to stop his helplessness from touching her. Her voice rose high and strained, "She will get better—she has to."

A knock came at the door, then it was pushed open by a man with a satchel. Her father nodded, "Come in, Doctor," he said, and the two hurried through the hall. The door opened again and the minister came in looking unearthly—like a black-clad angel of death. When her father came back, those two talked quietly. Mandy caught Pastor Palmer's quietly spoken words, "We brought nothing into this world and it is certain we will take nothing out."

Anger ripped through her, and like a strong wind turned her inside out. Her hands shot out towards the table where the boys were sitting. "My mother did bring something into this world—see, and she's not taking anything out—because she's not going." Then an overwhelming sense of helplessness seized her and she dropped to the daybed sobbing, "If Mom dies I'll hate God forever."

She jumped up and ran to her room with tears splashing down her face. The sudden sounds of laughter from outside the house hit her ears like a mockery. The rain had ended, the air was still and two girls in raincoats, one carrying a flashlight, skipped down the road as if nothing was happening—as if the world was not filling up with black emptiness.

Mandy tried to stop sniffling. She gulped back sobs as she went to her parents' bedroom. She wrung up a cloth and wiped her mother's face. "You're going to get better," she whispered. "You're just tired."

"I'm fair warped lying in this bed," her mother murmured. Then she smiled wanly. "I'll be like Lazarus. I was telling your father about my headstone. I saw my name."

"Don't, Mom, don't!" Mandy thought she'd die right then if her mother said any more.

Her mother's voice came strong. "A hand came, Mandy, a hand with a nail scar. It knocked the stone down."

New hope, like a living thing, turned in Mandy. She now knew that God didn't want her mother yet. She went to the bureau and pulled out the bottom drawer. There it was, the box with the pretty blue bed jacket her mother was saving for special. She lifted it from the folds of tissue paper and laid it across the bed. For tomorrow, she promised herself. Then she went back to her own room.

It wasn't easy to settle into sleep, and when she did Mandy felt a sudden tearing inside her as if a spike was forcing her insides open. Waves of pain made circles that widened and lifted her until she felt as if her body was tearing apart. Horses came tearing out of the wallpaper galloping through her head. The flowers in the hooked mat beside her bed leaped up and filled her mouth. She struggled to scream against the bed writhing beneath her. Voices floated past her in tunnels. She was being sucked into a dark hole. Suddenly a harsh light shone in her eyes. The sound of a scream caused her to spring up, a scream that seemed to be her own, then shifted to her parents' room. Her feet hit the cold canvassed floor before she realized what she was doing.

Mandy rushed to the closet and put her ear against the wall. There came a tiny cry, then an unnatural stillness. A voice said apologetically, "Too bad, he looked like a healthy boy."

Mandy crept into the hall and pushed at the bedroom door. It opened a crack and she saw Dr. Evans wrapping a baby in a blanket; he covered its face. Her mother's face drew up in pain. Her mouth opened, then closed as her teeth clamped down on her lip, forcing out a strained cry. Her hands held feebly to two strings of flour bags tied to the bed posts.

"My Lord!" exclaimed the doctor. "There's another one."

Mandy crept back to bed stunned. Another one! She strained to hear what was happening, but the voices were too low.

It seemed forever before she heard her father coming through the hall. She called, "Dad!" He came into the room and switched on the light. He was dressed as Mandy knew he would be, in his Sunday's best, and freshly shaven. He spruced up every time a new baby came along. That was his tradition. Now he looked more relaxed than he had in weeks.

He cleared his throat, "Your mother will have a better chance now; you have a sister." He turned to go, then reached back to touch her hair. His voice cracked, "I never dreamed I'd have a little girl like you with beautiful, silky hair and matching beauty spots." He left abruptly.

It was as if the sun had come back into the sky. Mandy pulled the blankets over her and settled down to sleep, feeling like a penned gull suddenly set free to ride the waves and air.

She awoke to see streamers of sunlight slanting across the floor. She tiptoed through the hall to her parents' room. The door was open, and Mandy saw her mother lying in the big feather bed. Her long black hair, free from its roller, lay across the pillow like a rippling black stream. She was wearing the blue bed jacket embroidered with thick flowers in pastel colours, and sequined with pearl buttons. Her father was holding a tiny baby, whose face puckered before letting out a sharp cry. He stroked her cheek, and she settled into making soft sucking sounds.

Mandy felt an overwhelming sense of wonder at the picture of her parents and sister.

The lady in blue finally had a face.

She wasn't really vain. How could she be? No one would let her—not her classmates who called her nicknames... nor her mother who cut a week off her summer stay with Aunt Callie because she smeared her lips with chuckley pear juice, and put gravy browning on her eyelashes. Not even God would let her be vain. He made her with red hair, a bad case of freckles and what the minister called a retrousse nose.

THE FEATHER

All of Mandy's longings seemed vain as she sat in church on Good Friday. A missionary had come to tell Missioners—Holy Rollers to outsiders—about souls in darkest Africa who had never come under the light of the Gospel, and had no hope of doing so if Christians didn't help. As she heard about mothers throwing their babies to crocodiles to please gods that could not see or hear, Mandy shuddered. When the missionary told about the Mau Maus stealing up over the hills of Kenya to attack missionaries, she was sure her hair would have stood on end if it hadn't been battened down in a plait.

Mandy would have given all her money to save the red and yellow, black and brown children if she'd been allowed to bring her purse to church. It would have been such a little sacrifice, she thought, considering that missionaries often had to eat insects and drink cows' blood. Some were sent to Heaven before their time by savages wanting to stay in darkness. But her mother knew that a girl's purse is a show of worldliness and distracts from worship, when it contains, not only money, but tools of vanity such as a looking glass and a comb.

Mandy felt relief, then guilt. She tried to defend herself. It had taken her from Pancake Night until now to save a dollar and twenty-five cents. That was the price of nylons she'd been longing to buy. She would have had them already if she had given in to the temptation to keep her Sunday School collection. She'd resisted. It was no good getting her first pair of nylons only to burn in Hell for keeping money that was meant to save the heathen in lands beyond the sea.

As long as she could remember, winter and summer, Mandy had worn brown ribbed, heavy stockings, held up by garters cut from white elastic and tied in a knot. Her father wouldn't allow her to wear ankle socks. Rubbing his big hairy hand over his balding head he'd told her, in an authoritative voice, that nakedness is an abomination unto the Lord. He also frowned on women who wore shorts, and women who showed their armpits. He was sure they had the spirit of Delilah or Jezebel in their hearts. Mandy knew he couldn't stop her from wearing brown nylons though—because they were much like the stockings her mother wore, only hers were grey lisle.

Aunt Callie was Church of England. She would never let on that she knew her sister thought she was on her way to Hell. Summertime, she wore thin nylons, showed her armpits, frizzled her hair and wore lots of jewellery. "All dolled up" was the way Mandy's mother described her sister, after she'd been for one of her visits wearing flashy brooches, and rhinestone necklaces and earrings, the signs of a worldly and unsaved condition. Once when her mother wasn't around Aunt Callie had dropped a string of pearls into Mandy's hand and warned her not to show them to her mother—yet. Then her aunt sighed, "Hopefully, she'll forget her nonsense some day and realize there's no vanity in looking nice. If Heaven's gates are made of pearl, and its walls of all kinds of jewels, I can't see the sin in getting used to a bit of it down here. Sure, 'tis only to make you look nice—a bit of decoration."

Mandy was lucky that Jeff didn't see the pearls. He'd tattle to her father. Ever since she got him into trouble after he'd built a scaffold and rigged up a tightrope, her brother had been poking her with his fists every chance he got. Her father tore the contraption down after he saw Mandy climb the scaffold. She'd just wanted to be up high so she could look across the hills and down on her house nestled in the hillside. Everything looked different from high up. It was as if the hills were cut out of a paper sky.

Jeff was considered to be in a *backslidden* condition after he was caught smoking tea leaves and going to dances. He'd even said damn in front of his father and tried to defend himself by saying Jesus said it. Lately, he'd been saying strange things about God. "God could have had a million sons. He didn't have to ask his only son to die. What did he want bloodshed for, anyway?" he asked defiantly. "He could have washed sins away, any way he wanted." Jeff never blasphemed in front of his father, but his mother heard him and went around with a grieved expression.

Mandy was glad when Saturday came. She could forget missionaries, Hell, vanity... everything but her stockings. As soon as all the house cleaning was done she hurried to Aunt Sara's shop.

When Mandy asked for a pair of nylons, the old woman looked at her over her wire-rimmed glasses, her grey eyebrows moved together, "Have you got the money?" Money was very important to Aunt Sara who was nobody's aunt, and nobody's mother.

Mandy shoved dimes, nickels and cents towards her. Aunt Sara counted them, then reached behind the counter and pulled up a box. She flipped through it and pulled out a pair of nylons on a card. She squinted her eyes at them. "I guess this is your size." She looked Mandy up and down, stopping at her heavy stockings. "Getting set to show off your legs are you, Girl? I always get you mixed up with your sisters."

"My sister is still a baby. Then there's Jeff, Michael, —"

Aunt Sara cut her short. "Michael," she gasped. Her eyes seemed to double in size and almost pop over her glasses. "My, that's a Catholic name, and you crowd are Missioners too. I'm Church of England and I wouldn't baptize a child of mine with that name. Your father was once Church of England and he knows that the Catholic church is drunk with the blood of Protestant martyrs."

"Michael," said Mandy defensively, "is not a Catholic name. It's in the Bible—and it's the name God gave to one of his archangels." Not that Michael was an angel. He got sulky because he wasn't going to get an Easter egg like other children in the cove whose parents weren't so strict.

He'd even answered his mother back when she was firm about it. "There's nothing wrong with Easter eggs," he cried, "the earth is an egg that God hatched everyone from."

Mandy didn't care about getting an egg. She had her nylons and she hurried home, rushing straight to her bedroom. Very carefully, she reached into the paper bag and brought out her first pair of nylons. They were so soft, so silky—and shaped like a leg she could see through. She made a fist and slid her arm gently down one stocking. It lay on her arm as light as a breath. She eased her hand out of the stocking and went to get her black garter belt. It had come in a box of clothing Aunt Ruth had sent from Grand Falls a few days before. She slipped it over her hips inside her long-legged cotton drawers, promising herself a pair of *step-ins* next. Her fist went down into the stocking and let the mouth of the stocking fall towards the foot. She slipped her toe in and eased the stocking over her leg, up over the ugly trench mark her garters had made, folded it on the knob of her suspender and fastened it in the front and back. In the same unhurried way, she slid on the other stocking. She couldn't believe it! Her legs were covered, and yet they looked naked. The red birthmark on her right ankle looked like a brown beauty mark and her legs felt so silky. She was tempted to slide her legs together, but she was afraid of hitching, afraid of getting a run.

"A run!" she cried, feeling a zigzag sensation down her leg. She watched in horror as a line ran straight to her ankle. It wasn't her fault, she decided quickly. The stockings were going back to the shop. She closed her eyes as if to shut out the sight of Aunt Sara. She could be mean. One cent owed her was recorded on a scrap of paper that ended up on a spike; and she was stingy enough to break a biscuit in half so she wouldn't give a customer more than the pound she was paying for. She tried not to think of these things as she pushed open the shop door and walked slowly across the wooden floor to the counter where Aunt Sara was busy counting the sweets in a large glass jar. One sweet rolled across the wooden floor in front of Mandy's feet. Aunt Sara hurried outside the counter and picked it up. She wiped it across her flour bag apron before counting it in. It was the same apron she used to wipe her knife on after cutting cheese, or salt beef.

Aunt Sara's stern face looked into hers, and Mandy faltered before pushing the bag across the counter. "There was a run in the stocking," she said quickly, "I'll have another pair."

Behind her glasses, Aunt Sara's eyes looked vexed. Her words were abrupt, "Nylons don't run on their own."

"These did," Mandy answered with conviction. Her longings pushed away her embarrassment and she added boldly, "I've just started buying nylons; these were my first. I won't be able to buy any more here if they have runs."

Perhaps fearing that Mandy might go up the road to Pleb's shop, Aunt Sara shook her head and said grudgingly, "This time, just this time." She took the nylons out of the bag and put another pair in. "Maybe I can sell them to someone whose eyes aren't as keen as yours." She stuck a gnarled finger in Mandy's face. "If 'tis not the truth you're telling, young lady, your legs—they'll roast in Hell some day. You mind now to put wet soap on a run when it starts. That way your stockings will last longer."

Mandy felt good as the door slammed behind her. For awhile she'd been afraid she'd have to go back to wearing her brown stockings. Now she could get ready for Easter Sunday.

When she got home she washed her hair in Halo, breathing in its fragrance, and rinsed it in vinegar and water. According to a magazine article she'd read, vinegar would make her hair shine. She was glad her bangs were finally long enough to put in kiss curls. She'd cut them in January. It was hard to get them straight, so she kept cutting one side and then the other. When she'd finished, they were sticking out from her hairline like the bristles on a paint brush. Her mother told her that she'd brought on her own punishment, and Jeff teased her about her brush cut until she'd warned him, "If you don't stop, I'll tell Dad about you cutting off your eyelashes just to see if they'd grow back."

She wasn't going to sleep on hard rollers with their bristles and picks tonight. She had started cutting strips from a biscuit box and doubling each one to roll her hair on when her father saw her. His voice carried as sharp an edge as the cliffs in the cove, "What are you doing that with your head for?"

Mandy's stomach went tight. "I'm going to bed right after," she promised.

"You better," he said darkly, "or that'll all come out."

It was like that with her father. He'd let her mother wear a big roller in her hair all day because she didn't have vain intentions. Mandy was allowed to sleep on her rollers, but her father wouldn't let her go around daytime looking like someone sitting in a beauty parlour.

" 'Twould be a better world if people left things natural," her father grunted as he lifted his razor to his face.

The sudden image of her father with his hair growing down his back, a long beard sweeping his knees and a mustache like his younger brother had, one that ended in curlicues up the sides of his nose, made her want to laugh. She didn't say anything to him. He would throw off any argument of hers as if it was air hitting brick. She walked past him on her way to bed, suddenly feeling a tension she couldn't name.

No matter how early in the year Easter Sunday came, it always seemed to bring spring. The light seemed to be brighter than on other mornings, and the air breezing through Mandy's open window was fresh and sweet. She was always stirred awake early on Easter Sunday morning by the chirping robins on the light pole wires outside her window.

Mandy wondered pensively if the sun really does dance on Easter Sunday morning, and would she see a cross if she held her handkerchief against it? She slipped back into sleep to be awakened by the VOCM church service.

Her mother, back from the sunrise service, was warming leftover fish and brewis when Mandy got up. There was a newness about the day. Perhaps it was the smell of her new shoes and nylons, or maybe it was because she was wearing spring clothes again. Her spring dress wasn't new, and her coat was last year's, turned inside out—to hide the faded outside. Her white straw hat was also last year's. It was shaped like a plate and trimmed with a blue grosgrain ribbon. For-get-me-nots and other flowers that looked like the bluebells that grew in the cliffs of the cove had been removed by her mother. She'd protested, "But the other girls wear flowers in their hats."

"They don't grow there and they don't belong there," reasoned her mother. "Beside, they make you look vainglorious. You can be well-dressed and wear a hat to worship the King without looking so worldly."

Mandy thought she could sneak out before her father saw her nylons, but she was too late. She was surprised that he merely said, his voice only mildly irate, "You'd never know your legs were covered if there wasn't a sew running down the back like a cut in a cod's belly."

The next time, Mandy decided, combing her fingers through the waves in her hair, I'll buy a pair of nylons with a black, velvet butterfly on the heel. She giggled at the thought that her father would probably spray her heel with fly killer.

On Easter Sunday, the left side of the church, where the men and boys sat, was dark and dull-looking, as usual. On the right side, where the women and children sat, it was as if a flower bed had suddenly bloomed. There were hats in all colours, shapes and sizes. Mandy and the preacher's daughter, who was wearing a white beehive, were the only two girls to have their flowers removed.

Pastor Palmer was quick to admonish his people that this was no day for pride to creep into the church, no time for saints to envy the Easter bonnet of another saint. He banged his fist on the pulpit, and roared, "The Lord has risen!" He didn't smile as if this was a happy occasion. To smile in church would have been to act frivolous.

Mandy tried to keep her mind off the feel of nylons on her legs. Only a couple of years ago, she completely forgot she was in church. There she was, chewing on her elastic hat string, and rubbing her spit-wet finger through the dust on her new, patent leather shoes. She'd marvelled at the shine the spit made. Suddenly she realized that Pastor Palmer wasn't talking. She lifted her head and met his gaze. His finger seemed to point at her as his voice came like thunder, "Vanity of vanities, all is vanity and vexation of spirit."

She wasn't really vain. How could she be? No one would let her—not her classmates who called her nicknames and chanted after her, "Hair from rust to dust, if God don't have you the devil must," nor her mother who cut a week off her summer stay with Aunt Callie because Mandy smeared her lips with chuckley pear juice, and put gravy browning on her eyelashes. Not even God would let her be vain. He made her with red hair, a bad case of freckles and what the minister called a retrousse nose. She pretended not to care about her freckles, though she was always looking for ways to get rid of them—short of scraping her skin off. Once when she asked her mother if black people would turn white if they washed their face in Javex, Jeff had taunted her, "That's not what she wants to know." He pointed to her freckles.

Aunt Callie had told her, with a twinkle in her eye, "Drink lots of milk and your freckles will turn white. Then your cat can lick them off."

"White freckles!" she laughed, feeling relieved that she didn't have them.

Once the singing got started, all the stops were pulled out and the little church vibrated. Before long, Mandy forgot about the silky rustling sensation of her nylons. A soft peaceful feeling filled her insides, swaddling her heart in its softness. She joined in, clapping her hands until they burned and stung from the friction. Her heart danced with the triumph of the resurrection.

If only her mother was there to feel it, instead of at home cooking dinner. She'd watched her mother in church sitting quietly, tears rolling down her face. She saw her worn hand wipe the tears away with her white handkerchief, her face shining as if it were oiled. Her mother, Mandy thought, loved God more than anything, more than anyone; she lived only for The Rapture. It was hard for Mandy to have a mind for planning the future when no one around her expected to be here on earth. Any moment, in a twinkling of an eye, she could be gone.

Aunt Edith was testifying, saying the same things she said every Sunday. She seemed to get blessed talking about the wonderful mansion God was preparing for her in Heaven. She flipped and flopped around like an old washing machine. This world was not her home. She was going to Heaven where she would wear a golden crown and walk on streets of gold. It was easy to see she took great pride in having her hair rolled tight to her scruff under her plain grey hat. Mandy would never forget the hurt look in her mother's eyes when Aunt Edith told her that if she had a son like Jeff she'd never darken the church doors. That was after he played his mouth organ during a young people's service. Mandy couldn't imagine what she would have said if she'd known he'd been caught smoking tea leaves and going to dances. Her mother didn't say a word, not even when she got home. But when Aunt Edith's Julie, who was Jeff's age, got in trouble, Jeff joked, "She believes in 'doing it' the Missioners way, 'cause if she 'did it' standing up, it could lead to dancing."

Uncle Pleb was next. He started pacing across the front of the altar, his red-rimmed eyes looking out under dark, bushy eyebrows. He blew his nose, then waved his handkerchief and shouted, "Hold the fort for I am coming." People started to sing, and the organist began playing. When the hymn was sung, Uncle Pleb was still standing. He pointed to his heart. "My friends, I'm glad I was simple-minded enough to give my black heart to the Lord." Mandy used to think that hearts were either black or white—never red—until she saw a picture of one in Jeff's science book.

When Pastor Palmer began his sermon, he warned that saints who were encumbered with the things of the world would miss The Rapture. "Let us lay off the sins which doth so easily beset us," he shouted. "Earrings are stirrups for the devil to ride in and necklaces —"

Mandy didn't wait to hear more. She grabbed the pearls hidden inside her dress before they could become a weight that could keep her from Heaven. She didn't want to be left to take the mark of the beast in her hand or forehead like cattle being branded. Sometimes she could imagine herself being left behind, to hang on the light pole beside her house by the hair of her head for refusing the mark. If she didn't refuse the mark, she still wouldn't live happily ever after. Hell was waiting for traitors of The Gospel.

Without thinking, she pulled hard on the pearls. They snapped; down they went, rolling across the wooden floor. Some stopped by Aunt Edith's shoes. She turned to look at Mandy, her lips pressed tightly against her teeth. All of Mandy's heavenly feelings left her. The humiliation grew worse as she tried to hold some of the pearls inside her dress. As they continued to drop, the heat rose in her face, and stayed there until the service ended.

She pulled the nickel her mother had given her for collection from the finger of her white lace glove and dropped it into the box on the door, then hurried outside and down the steps. In her embarrassment, she forgot about her nylons. Then all of a sudden, it happened. Someone pushed against her, and she missed the last step. She fell on her knees to the ground. The feel of sharp stones against her knees didn't bother her. All she could think of was her nylons. As she got up, she felt the sensation of the stocking letting go. Tears filled her eyes as a host of runs shot down her leg. Her nylons were ruined. She looked up quickly as Aunt Edith came down the steps. She shook her head haughtily and quoted the Bible, "Pride goeth before a fall." Then she went on by without asking if Mandy was hurt.

It's a pity, Mandy thought angrily, she can see so well. She wouldn't if Pastor Palmer hadn't suggested to her that glasses don't make people look vain. He even dared to suggest that God may have given people the outer part of their ears to carry their eyeglasses. Now Aunt Edith made sure she used the parts of her ears to hear—and see.

When Mandy got home, she ran past her mother's questioning look into her bedroom. She shut the door and lifted the nylon off the torn skin. It was stuck on with dried blood and mixed with dirt. It was too late to save her stockings, so she let her dress fall over her knee and came out to dinner.

As she ate the Jiggs dinner: vegetables, pease-pudding and salt beef, she felt better. But as she helped clear away the dishes, she was thinking of ways to get another pair of nylons. She was more determined than ever as she pulled off the nylons and put the cocoa-coloured, thick-ribbed stockings back on before going to Sunday School.

After Sunday School, Mandy and her brothers came home to an early supper. As soon as it was over Mandy turned on the radio to enjoy Your Story Hour, a half-hour program of Christian plays that she listened to almost every Sunday. Her father had gone for a nap before Sunday evening prayers, and Jeff decided he was going to turn the radio down low and listen to a hockey game. It was something his father wouldn't allow. He believed that if the radio was used for anything but the news of the day or the news of the gospel, it became a tool of the devil.

Jeff was determined. He turned the dial and Mandy got up from sitting under the shelf the radio was on and turned it back. Her mother always said she was stubborn. Now her stubbornness lay inside her as hard as a walnut. It seemed she was always fighting the world, the flesh and her brother.

Out of the corner of her eye she caught a glimpse of something in Jeff's raised hand. It was one of a pair of spike-heeled shoes Melva, her cousin from Gander, had left after a visit. The heel was three inches high and had a metal tip. When it hit her head, she crumbled to the floor in tears. Her hand went to her head. It came away sticky. She ran screaming through the hall into her room, "My head! My head!" All the time she was thinking: What if my brain is stabbed! It will stop working, and I'll die and go to Hell for fighting with my brother.

She heard her father's heavy footsteps moving faster and faster through the hall. He was pulling his belt off as he went. The clink of the buckle sounded harshly in her ears as Jeff's face came to her—Jeff's face with the defenceless look he got in his brown eyes when he knew he was in trouble. She stopped sobbing long enough to listen to Jeff's angry groans. He hardly ever cried now that he was older. Suddenly it seemed all her fault.

Her mother came to examine her cut, and wrap a towel around her head. "Hush!" she chided gently. "The cut is deep, but small. If you'll stop sniffling and settle down, the blood won't flow so fast."

She tightened the towel around Mandy's head. "Keep this on for the night. We'll wash your hair in the morning."

"I have to wash the blood away before then," she sobbed.

"You can't," her mother said firmly. "You know your father won't allow you to wash your hair on Sunday."

Her mother went out and Mandy closed her bedroom door. She flopped down on the floor of her closet. It was where she always went when Jeff hit her. She spied a Harlequin book sticking out from the corner of the shelf above her head. She got up to push it out of sight, afraid that her father would discover it. Then she sat down again on boards that had books and other things hidden under them.

There were so many things her father wouldn't let her do. She sat in the closet thinking things she dared not say for fear her father would rebuke her. Sometimes it was as if she was made of glass, and her father and her heavenly father could see right through her. The lidless eye of God shining into her soul, seemed as much a danger as the Khrushchev she'd read about in her father's *Reader's Digest*. The Russian people had to watch their step around him, but, at least, he couldn't see inside them; a lot Russians didn't have a preacher to tell them that there was someone who could.

Mandy could hear movement in the next room. Her parents were getting ready for church. Her mother's hair would be let down her back and she'd be combing *clits* out of it. Then Mandy's father would take the electric wire he'd cut for her roller, slip her hair between it, and roll it up to her shoulders. Her mother would take it from there, bend the ends in under and shape the roll like a sausage in the nape of her neck. Then she'd pull a dark hairnet over her hair to smooth stray hairs in place.

Mandy came out of the closet in time to see her mother dressed for church. Her mouth dropped open at the sight of her mother. She looked positively vain—as vain as a peacock, Mandy thought, then wondered why anyone said that. If peacocks looked vain it was because God made them that way. It was humans who were always changing their looks. Her mother was wearing a white felt hat with a blue feather coming up around one side. The rim was swept up on the same side as the feather. The hat fitted her head perfectly. Beneath the dark hair waved across her forehead, her eyebrows arched like wings. Her deep-set blue eyes looked bright as they met Mandy's. She couldn't help noticing that her mother's high cheek bones were flushed as if they'd been rouged; her lips were soft and full and as red as if the colour had been bitten into them.

She'd never before thought of her mother as vain. Elizabeth always put on her hat with hardly a glance in the looking-glass to see that it was straight. Mandy knew though by the way her mother studied the hats in Eaton's catalogue, that hats were her weakness. Not that she bought a lot. Most of them came from her sisters in Grand Falls. One summer, when the fishing was good, she'd sent for a hat in Eaton's catalogue. She'd lifted it carefully out of the tissue paper and box it came in and pressed it down on her head. Mandy's father came into the room just then, and asked—as if it was an old Cape Ann—"Where'd you get THAT?"

"I bought it," Elizabeth answered quietly, slowly lifting the hat from her head.

"I don't know why you did that," he said flatly.

"You don't?" Her mother's voice was strained. Then she put the hat away. Mandy knew why. Her mother thought it was only fair to her husband to wear something he liked. After all, he was the one who was going to look at her in it. She'd see herself in it only while she was looking in the glass.

Mandy didn't mean to mention the feather. She wouldn't have if she hadn't thought of all the pretty flowers that were taken out of her hats. Her words were impulsive, "You can't wear that feather, Mom. Feathers don't go in people's hats. They belong to birds." Mandy saw something come into her mother's eyes, a sudden recognition of her own vanity—her sin. The awareness brought a pitiful look to her eyes, one Mandy would not forget as she realized that her mother would never forgive herself for her show of vanity. Without saying a word, she pulled out her hat pins and lifted the hat off her head.

Mandy felt her mother's silence fill her own insides like a cold stone as her mother moved away from her and went into the bedroom. She wanted to call after her, "You didn't have to take out my flowers, and you don't have to take out your feather." But she couldn't say a word. She pulled the towel tightly around her head to keep the blood-stained strands of hair from falling into her eyes.

When her mother came out into the hall the feather was gone. The hat looked bare and plain, and so did her mother's face. Her father and mother went off to church with her four brothers. Mandy settled Elizabeth into her cot, then she went back to her room.

She lay on her bed feeling the weight of a religion she could not understand, one that seemed to take over her life and fill it with conflict. She remembered seeing a blood-stained chick peck its way out of its shell. She wished she could peck her way out of a tight, dark world. Perhaps she could let her thoughts go free, instead of bracing them in like her mother braced herself in her corsets.

She was glad she had not been born in Africa where heathens believed stone gods could see and hear, and were everywhere looking into a person's soul. Perhaps she should have been born on Easter Island before the Christians came. Then people made statues; they knew they were only statues. They didn't try to make God in their own image.

All her life Mandy had to believe in a God defined by men she never knew, men who had done a lot worse than be vain and act natural. Some of them had stoned people in God's name. She wished everything she'd heard about God would disappear, wished she could hear him speak for himself.

Her mother said that God speaks in the thunder. He uses it to warn the wicked that they are sinful, and his eyes are in the lightning, signalling people to fall on their knees in repentance.

She sat up and pulled the heavy towel away from her head. It was covered in blood, and her hair was beginning to feel crusty. She wasn't going to sleep like that. She went to the bathroom and turned on the cold water. Then she went to the stove and got a dipper of water from the hot water tank. As she put her head down in the sink, the water turned crimson. It finally cleared after she rinsed her hair a few times. She wrapped her head in a clean towel, feeling clean and glad that she'd done something she wasn't supposed to, something that was right.

Mandy did not kneel by her bed and cast her sins into the sea of God's forgetfulness as she had done other nights. She slid between the comfort of flannelette sheets and pulled her daisy chain quilt up tight to her chin. Knowing only God could hear her, she said aloud, defiantly, "I'm not going to believe in you anymore—not the way other people tell me to. I'll let you speak for yourself."

She was just dozing off when, in the distance, she heard the sound of thunder. One clap followed another, getting closer each time. The sounds were like barrels rolling across the sky. It was almost as if the sky would break in pieces and fall on the house. A flash of lightning lit up her window.

Mandy scrambled out of bed, falling on her hands and knees to the cold canvas floor. She crawled into the closet and began to pray for silence.

She would go back—sometime—to cleanse the spot and wrinkle from her soul, but first she had to think about all the things that might happen to her if she did. She could be put in jail. She imagined herself bringing in her thirty cents, and, while the cashier was busy, laying them on the counter, then leaving.

THE DEVIL DANCES IN EMPTY POCKETS

Mandy reached into her closet and felt inside the pocket of her white dressing gown. Her heart started to beat in panic until she touched a hole. They must have dropped down into the lining, she thought. She sighed in relief as her fingers felt something long and round, then another. She sniffed the air, afraid the smell might go through her closet, even though she'd hung a cake of Johnson's soap from the hanger. She had no choice but to hide cigarettes for Jeff. She had done a terrible thing and he had threatened to tell on her if she didn't do what he demanded. "No one would know," Jeff said. He'd run out of hiding places in his own room. More than once his mother had found things he had hidden. Mandy knew she was paying the wages of her sins by being caught forever in the torments of her older brother.

It was only last Sunday that Aunt Jane, her Sunday School teacher, had warned, "People who steal things are as bad as murderers." As she spoke, she was looking right at Mandy, almost as if she knew her transgression. "They will burn in Hell." She shook her head at the pain of it. Her shoulders tightened and she leaned towards Mandy, who couldn't help noticing her breasts hanging like a britchin taken from a large codfish.

Mandy tried to imagine burning in Hell. Perhaps the only time she had understood the absolute pain of it was when she was melting crayons on the coal stove to make little scabs and she'd toasted the tip of her finger. That was Hell!

"There are words to describe people who take things, words that make nicknames seem mild," Aunt Jane pointed out. Then, to illustrate how good and wise children can be, she told the story of the little girl who went to a lady's house hoping to sell her some partridgeberries she'd picked. She gave the lady her flour bag of berries and told her to measure out what she wanted. "I only took what I was willing to buy," smiled the lady, returning the bag.

"I knew you would," said the wise young girl. "If you stole some of my partridgeberries, you wouldn't get rich, and I wouldn't get poor, but you would be a thief."

A thief! The words seemed to be stamped on Mandy's forehead in red letters. Her head felt so leaden with guilt, she wanted to let it drop. She was sure it was showing in her eyes. With all her might, and everything else it took, she straightened her head and looked at Aunt Jane who she was sure had never done anything bad in her life. Unless you could call looking mean an act of badness.

The black spot had probably been in Mandy's conscience ever since she was born "in sin and shapened in iniquity." She'd known she was a thief from the time she went ice fishing, and did something so mean her brother, Taylor, never forgave her. She saw a trout he had caught flapping on the ice and while he was gone to another ice hole, she put the trout on her hook and dropped it down into the dark waters under the ice. She put her face down to the hole, her lips dipping into the cold water, and squinted her eyes so she could see the trout on her hook. When she pulled the trout up, her first impulse was to run for home. With each step the trout became more and more hers. She began to forget she hadn't caught it, and she was proudly holding the trout in front of her mother's eyes when all of a sudden she heard someone bawling. Taylor's voice, sharp against her ears, broke into her dream of having her first trout. He was accusing her of stealing his. She was defiant, unable to adjust to her disappointment. The trout had been hers all the while she was pulling it out of the hole, all the while she was running across the ice, all the while ... It was hers until Taylor's crying caught her ear and broke the spell, then it was Taylor's again. She didn't know why her imagination carried her so far—so far from what she was doing that she couldn't think of it as stealing. It was hard to explain.

"Taylor didn't own the trout," she shouted. "He stole it from its mother."

"Now," her mother said sternly, "you may as well pull the devil by the tail and confess. I know your look when you've been into mischief and the stubbornness that holds you to it."

Then there was the airplane sharpener. She'd never before seen a pencil sharpener shaped like an airplane, the colour of a flamingo. She'd gone into Robert's house to get warm after skating on the pond, and had seen the sharpener that Robert's relatives had sent from Toronto. He didn't seem to care about it, and she had room in her leggings. Once she got it in her hand, it seemed right to drop it into her pocket. It promptly slid through a hole down her leggings and stopped inside her boot at her instep. Her conscience hit immediately, and she wished she could put the sharpener back on the vinyl tablecloth before someone saw her. She went home and dropped it on the floor behind her bed.

"Can't you call this my learning time?" she had pleaded with her father when she saw his leather belt and heard it make a snapping sound as it left her father's trousers. "How am I ever going to learn not to steal if I never feel the hurt of my conscience?" she had sobbed. The cutting of the belt in red welts across her hands and legs only made her angry.

Then there was the bar. She'd been really hungry that day when she'd gone to the shop for her mother. Her mother had wanted her to charge only the bare necessities. "The bar too?" Aunt Sara had asked, following her eyes to the candy bar on the inside shelf.

"Yes, please," Mandy had answered quickly, knowing that Aunt Sara didn't care what she charged. Her mother would pay for it as soon as the first haul of fish brought money in the spring.

As she trudged home, the load of groceries made her arms ache unbearably. She went in and sat down on the mission steps, dropping her parcels on the ground. She opened her bar, and sat for a long time sucking chocolate off lots of nuts. Why was she eating a forbidden bar on sacred ground? she wondered, and then kept right on eating. Perhaps she had a bad conscience.

Mandy's mother had looked on the bill. Her dark eyebrows moved towards each other. "Did you charge a candy bar?" she asked. Mandy wanted to reply with the words her mother used when she asked her something she didn't want to answer: "Ask me no questions and I'll tell you no lies." Instead, she said quickly, "No." The "no" filled the air like a fire brand. Suppose she died with a lie in her throat? She could imagine the pond by her door becoming "a lake of fire" just for her.

"Then," asked her mother in a civil voice, "why is there a bar charged on this bill?" Before Mandy could answer, she continued, "Be sure your sins will find you out." Her mother always used the Bible as her support, and, fortunately, considered Mandy's awareness that she was a transgressor punishment enough.

Mandy and Janice, who had been best friends for a full year, liked to go to the post office after school. They often dropped in to Jacob's store if they had a dime or even a nickel. They hurried up the wooden steps into the warmth of the old shop. Inside it was big with wide wooden floors and a counter on each side. At the side door sat retired fishermen, yarning and smoking pipes. Towards the back, there were stairs leading to an upstairs room where the leather smell of new shoes and boots filled the air. Sometimes the storekeeper's wife, Elsie, would open the door and come in. She was tall and stately-looking, made from money, it seemed, and, as an old fisherman remarked with a squinted look, "as streamlined as a ship."

Mandy and Janice were only interested in the main part of the store where candy jars lined the right-hand side, candy jars that took two hands to open, all six of them with myriad candy, all shapes and sizes and tastes. Either ten cents—or five—could buy a whole bag if Maryannie was serving that day, but there would be room enough to fold the bag over if the storekeeper's wife was serving. Some of the candies were four for one cent, while others were one cent each, but Maryannie would just ask if she wanted some of each, and Mandy would nod and try not to feel guilty. She wouldn't look at what was put in the bag; but, as her hand closed over it, she knew she was taking more than she was paying for. Still, she and Janice would giggle as they went down the road, their mouths full.

Mandy often went to Bayview with her father and Jeff, but this day was one she would never forget. She was now branded as a thief forever, because this time she was old enough to think before she acted and to have a good conscience formed. While her father went for boat supplies, she ran to the drug store. Mandy hoped to be back to the truck before her father discovered that she'd bought Harlequin romances. He didn't approve of her buying any love stories, believing that thirteen was too young an age for girls to think about such things. As she stood looking at magazines and books in the book stand, she fingered the two fifty cent pieces in her pocket. She could buy three books—or—the thought occurred to her that she could prove to her brother that she could get away with whatever he could. She enjoyed reading books he had stolen, although they were violent and hard to understand and there was a lot of sex talk in them. The books were all in a suitcase that used to have a naked girl inside the cover until she'd torn the picture down because it made her feel as if the girl's body was hers, and it was she who was naked.

She lifted a slender book from the stand. "Runaway Nurse" stood out in bold letters. She could have bought it. She cringed as she slipped the slender book into the elephant ear pocket of her lime green coat. Then she picked up two more books and paid for them. When the door slammed behind her, she felt a sense of surprise that nothing had happened and then a swift moment of triumph. Suddenly her legs went rubbery and she almost tripped over the step. What if she had fallen and the book had been hurled right out of her pocket? Her triumph exploded in fear. She was enveloped by emotions so intense she wanted to disappear. She expected a policeman to run and grab her from behind. She'd have to go to jail. Her father would kill her if that happened—ruining his good name! A terrifying sense of danger surrounded her, filled her and drove her as if she was between a wind and a fire. Her footsteps seemed to echo what she had done. She couldn't stop trembling as she hurried faster and faster towards her father's truck. One hand was in her pocket on the stolen book in case she tripped and it fell out of her pocket and the other hand was on the books she wasn't allowed to buy. She was enveloped by emotions that filled her until she thought she'd burst with the weight. She finally got inside the cab of the truck and slammed the door. To her relief, only Jeff was in sight.

"I did it," she whispered into her brother's ear, smiling to make her feelings settle down. "I stole a book—just like you." She wanted to prove to her brother that she could do what he could do even if she hadn't conquered walking on his rope, and rubbing the back of her head with her toe. It hadn't occurred to her that there was pain involved—that her conscience was not seared like her brother's. He'd been transgressing for years. Even the youth leader at the church had given up on him after he'd played a mouth organ during a hymn. He'd been kicked out of school officially by his teacher, who'd chased him out the door after he'd screwed up his face at her. He'd protested that he wasn't making faces at her; his nose was being tickled by a sneeze. Miss Primmer slipped on a rock she'd crossed trying to catch Jeff and tore a hole in her stocking. She'd limped back to school.

Jeff turned abruptly, and hissed, "You didn't!" She felt as if she'd been slapped or dug in the ribs, like the time she'd laughed aloud when she noticed that Jeff had cut off his eyelashes. Now he was acting as if she'd committed a crime. She'd done only once what he had boasted he'd done a dozen times.

Now he was denying that he'd ever stolen a book. Perhaps Aunt Jane was right when she'd said, "As scarce as truth is, the supply is greater than the demand." She didn't want to hear that Jeff hadn't stolen books, now that she had stolen one. "What are brothers for?" she sniffed to herself.

She was glad to get home so she could hide her books until bedtime. Later she settled into bed, her toes digging deep into the flannelette sheets as she began reading the stolen book, stopping now and then to think about the other two she had left to read. Only now there was an uneasy feeling about the book she'd stolen as she tried to concentrate on Brian who was about to kiss Jill. She stopped, and listened. Was that her father's footstep in the hall? She tensed, but read on: *Their lips met*. It was her father's step. She waited for the clip to be pushed into the lock and the knob turned. She was poised to slip the book under her pillow. Luckily the hall light concealed her bedroom light. The switch was beside her bed so she could quickly turn off her light if she had to. She had hung a quilt over the closet that ran from her bedroom to her parents' room so they couldn't see her light on. To her relief, her father went past her door. She read on hurriedly and in misery, rubbing spit on her eyelids to soothe them when they began burning, trying to push the guilt from her with every page. But it was there pushing in at her, staining her heart black. By the time she'd finished reading the book, she felt exhausted.

She would go back—sometime—to cleanse the spot and wrinkle from her soul, but first she had to think about all the things that might happen to her if she did. She could be put in jail. She imagined herself bringing in her thirty cents and, while the cashier was busy, laying them on the counter, then leaving. What if someone saw her and asked why she left the money? If she lied it would be as bad as stealing. Perhaps she could leave her money where the book was in the stand. That would be putting temptation in the way of other kids. She wished she could bring the book back, but it was used now, anyway. Then suddenly it dawned on her. She would give it to Janice, without telling her where it came from. It was a momentary relief.

She pulled out a bureau drawer and stood the book on the ledge, along with books her father forbade her to read. She leaned back. She could go to Spain, Japan, Russia, anywhere. No matter that she was in her own bed in her father's house where everything that was done had to meet with his approval.

She avoided the drug store for weeks, but then one day she decided to go in. She would have to tell the manager—even if it meant going to jail. She had to break the habit before it took hold of her. The manager came out quickly and she held out fifty cents. "I stole a book," she said awkwardly, then added in a flustered voice. "I'm sorry, and now I'm paying twenty cents extra."

The manager looked at her as if he'd never seen a teenager before. "Well that's a first," he said in a voice that didn't seem angry or put out. "I wish more young people who shoplift would make restitution. You're a brave young lady—a good girl."

The word 'shoplift' danced in her head. It didn't sound as bad as 'steal', or 'theft'. She didn't feel so terrible, and her parents would never have to know what she had done if she did whatever Jeff wanted. That was the price she had to pay for having a conscience that wasn't always in working order.

She thought she paid her dues the day a salesman showed up while her mother was gone to the fish flakes. When she saw what the man was peddling, she felt like someone hungry. It wasn't aftershave lotion or kitchen cleaners like some of the other salesmen peddled, and her mother bought to "help the poor man out." The Fuller Brush man was popular. He had the dashing good looks of Clark Gable, and unlike the gruff voice of many fishermen, his voice was smooth. He enunciated every word, wrapping his products in such ear-pleasing sounds that if Mandy had been able to buy his words, for future replay, she would have. Her mother, she suspected, bought aftershave lotion like the salesman was wearing to help her forget about the fish smells Mandy's father brought home.

This man was a real author. His name was P. J. Wakeham. She remembered him from last year when he sold her mother *Princess Sheila*. In the book there were names of places Mandy had actually been. It was the story of a heroine who had lived not many miles away in the long ago years of the 1700s. Sheila Na Geira had been captured by the charms of pirate Gilbert Pike. The two were married and settled in Bristol's Hope. There the Irish Princess lived: fighting off savages, working miracles, losing her husband and getting him back. She lived happily until she was 105 when she passed into the pages of history—or legend.

Mandy eyed the title of the book in Mr. Wakeham's hand *Royal Imposter*. Her words, "Mom isn't home," came out heavy with disappointment. "But she should be back soon," she added hopefully, "once you've covered all the cove."

She watched him go, waiting patiently for her mother. She was reluctant to spend her two-dollar bill. That's what he'd charged last year. She went to get it, tucking it into her palm just in case the author returned before her mother. Then she went about tidying the house. She swept the floor, dropping the dirt from the dustpan under the damper of the stove, all the while thinking about the book she would soon be reading. She lit the paper in the stove. Just then she heard footsteps on the gravelled path. The author had returned—and her mother hadn't. She'd have to spend her money, instead of keeping it to buy a Salle Lee Bell book in St. John's. Suddenly she realized her hand was empty. Her hopes of buying the book sank as she lifted the damper and saw the black crisp paper of the bill she had unknowingly dropped into the stove.

Mandy shook her head when the author asked if her mother was home yet. She could hardly speak through her disappointment.

"Surely you have money," he said bluntly, holding the book up so she could sniff its newness as overpowering as books at Dick's & Co. in St. John's. She could only shake her head miserably and watch as he went around the corner of the house. For one mad moment she wanted to call out to him. She wanted to ask if he'd let her have a book if she gave him all the caplin that lay on the netted, hand flake like brown sticks, their eyes like little knots. But last year her mother had given him caplin for free. There was a salmon in the fridge. She couldn't....

She watched as he grabbed a handful of caplin on the way to his car.

She closed the door just in time for him not to hear the sob that suddenly escaped her. It triggered more sobs and she burst into tears, her body shaking with the agony of losing her money and a book at the same time.

She stopped crying, and her eyes brightened, as the thought occurred to her that she'd paid for everything now: the trout, the sharpener, the book. She wasn't going to Hell. She started to smile.

Suddenly she heard her mother's heavy footsteps scuffing up the gravelled path. Mandy's smile disappeared as she heard her exclaim, "What in the world is that in the hem of Mandy's dressing gown?"

Mandy ran to the kitchen and looked through the window to the clothes line where dark stains dotted the hem of her white dressing gown that her mother must have washed earlier that day.

Her eyes widened in horror. "Jeff's cigarettes!" she gasped.

She pushed the chair back unsteadily and, holding her covers tight to her chest, stood up. A kind of mesmerization took over her body as she moved like rubber towards the teacher's desk. She held her precious pictures out hesitantly, careful not to touch Mr. Rachit's clammy hands.

TORN PICTURES

Elene was Mandy's new classmate, the daughter of a teacher who had 'come from away'. As a fisherman's daughter who had lived in Maley's Cove all her life, Mandy didn't feel she had much in common with Elene, especially since Elene was, what the cove people called, a picture. Her silky Midas-touched hair had a springy look that came from having been bound in curls every night of her life. Mandy's hair was long and heavy, without a kink, and lately she had to put up with the name Coppertop. Elene did not have even one freckle like those splattered over Mandy's face. Her face was round and her skin as flawless as the face of the porcelain doll she'd brought with her from Toronto. Not even the snow in May nor the jellyfish in July had worked on what Mandy considered her second grade flesh.

There was no doubt in Mandy's mind that Elene's parents thought that their daughter was a notch above the cove children, not after she heard Mr. Rachit call Elene for supper. As she came around the corner of the two-storey house they were renting, he exclaimed, "Oh—is that you, Elene, dear? I thought you were one of the common girls."

Mandy had put her hand against her mouth to stop her laugh, but it burst out between her fingers. Still, she was willing to acknowledge a certain refinement about the tall, sissy-looking man who was always dressed like a stick of chewing gum. Her father dressed in his Sunday clothes mostly on Sundays. The rest of the week he wore thick plaid shirts and sloppy, faded overalls. Her mother too wore ordinary clothes during the week, and not even her Sunday's best could make her look like Mrs. Rachit in her expensive clothes. Mandy's mother wasn't very interested in reading about the rest of the world either. She didn't intend to travel far—not in this world. She was going to Heaven, and she found all she needed to know about that place in her Bible.

Mandy was always interested in finding out about the rest of the world. Ever since she was old enough to read the ads in the *Family Herald*, she had sent for anything offered free. Her collection included everything from postage stamps to religious pamphlets from Protestants—and Catholics.

Then one day, it came suddenly—the revelation that she was throwing away something wonderful that was also free: pictures on her scribbler covers. It was one of a little black girl in her frilly, white dress that first made her aware that she should start a new collection. The little girl was wearing a white, plate-shaped hat decorated with posies. She held a basket of flowers in her white gloved hands.

Other pupils in Mandy's class were willing to give her their covers once their scribblers were used up. Soon she had what she called her cover collection: twenty beautiful pictures, some of them copies of very old paintings.

Dora, who wanted to be Mandy's best friend, was leaning from her desk across the aisle to Mandy with another cover when Elene sauntered by. She *sliered* at Mandy, then smiled jauntily at Dora, "I'll pay you for your covers."

"Pay me!" Dora exclaimed, pushing a wilted-looking strand of dirty blond hair back from her pale face. "You'll pay me for covers

I'd throw in the garbage if Mandy didn't want them." The cover fell to the floor.

Elene stooped to pick it up. Dora looked uncomfortable. She couldn't meet Mandy's eyes as her hand closed over the dime Elene pressed into her palm.

Mandy didn't say anything. Her friendship with Dora wasn't so strong that it couldn't be cut from its moorings. Elene could buy what she wanted. That's what came from having a mother who painted pictures and sold them to an art studio in New York, and a father who was the teacher of grades seven to eleven in the cove's three-room school. Fishermen's daughters like Mandy and Dora had little money to spend during the winter months and early spring.

Mandy tried to get it through her head that she must not feel betrayed by Dora or any other pupil who sold covers to Elene. That was especially hard when there was a breathtaking picture like the one Dora got on a scribbler as a Christmas gift from her aunt in "The States." There sat a little girl with fat ringlets, her pudgy arms wrapped around a white lamb.

Cover after cover kept going to Elene and her collection got thicker and thicker. Mandy began to feel like someone hungry all the time. Her mouth watered every time she saw Elene buying a cover she didn't have—and wouldn't have because there was only one like it. Elene had gained on her and something that had began as a collection had turned into a competition. Mandy felt uncomfortable every time her blue eyes met Elene's green ones. She knew Elene didn't care about the covers.

It was a warm spring day, and Mandy tried not to think about covers as she walked across the path to her grandmother's home. She stopped on the hill before she got to the large, white house and let the wind lift her hair as gently as a mother's hand, as she looked out to sea, trying to see beyond the dome of sky that shut her in. She drew in the tangy air and let it out with a satisfied sigh. Soon the grass would be green, the bluebells in the cliffs would wave their heads, and the boy's love bush and honeysuckle scrub would climb the rock wall that her grandmother was sitting on now. As always, she sat as straight as a ramrod, her hair, inside her hairnet, braided at the back of her neck like a mat made from her mother's shades of grey lisle stocking. Her hands were clasped in her lap over her flour bag apron with wide pockets that usually held a mixture of buttons, pins, and peppermint sweets. She called to Mandy just as she lifted the latch on the gate and started up the gravelled lane. "Seeing how your nose is always stuck in books every chance you get, I thought you would like to look through the lot I came across in the shed."

The rocks skittered under Mandy's feet as she ran up the lane. Her grandmother laughed, "You don't have to *scravel*; the books will be there for a while yet, at least until I do my spring cleaning."

Wave after wave of joy swept through Mandy as she dropped to her knees in reverence beside a crate of books and papers. She lifted out a *Reader's Digest*, and a water-stained book of poetry bound in leather. Then she saw the scribblers! She had to close her eyes to shut out everything but her happiness at seeing the cover on a scribbler she'd never seen before, one that had her father's name on it in the scribbled printing that matched her own scrawl. The cover showed girls sitting in a carriage, all dressed like Scarlett O'Hara. She let herself lift that scribbler, then another slowly, feeling the thrill of her discovery swell and break inside her.

Her heart leaped as she struggled to get up under the weight of her treasure. Now she had the edge over Elene. That knowledge spread through her with such pleasure that she swayed.

Her grandmother shook her head, "That crate is too heavy for a girl as slight as you, especially going up those hills."

"But I can lift it, Gran," she answered breathlessly, her face burning with the heat of her excitement. "If my arms ache I'll just take a spell."

But as she braked herself going down the lane, and pushed herself across the path and up over the hills, her arms didn't ache. She didn't need to stop for a spell because she was thinking the whole time, and when she was thinking she was inside herself, away from her own tiredness. Sometimes on her way somewhere she imagined a whole story, unheeding her mother's warning that imagining things was the same as telling lies.

Only when she plopped the crate down on the steps outside her house did she feel a pinched ache in her arms from carrying the books. She stood looking towards the pond watching trouters cast their lines, breaking the smooth surface of water. There were shouts of glee when a pole bent under the weight of a trout on the hook. Mandy's mind turned back to her covers: one two, three...

Elene's eyes widened and they seemed to turn greener than ever before, when she saw the covers Mandy's grandmother had given her. Her eyes narrowed; she sat at her desk for a long time, not saying a word, her mouth like lip-shaped balloons. Mandy imagined sticking a pin in them, and letting them bang against her teeth like bubble gum.

Suddenly Elene turned her warmest smile at Mandy and held out her hand. She opened it and a dollar fell on Mandy's desk. "I'll give you this dollar for the picture of the kids on the cliff and the angel with the big wings," she offered.

A dollar—a whole dollar, thought Mandy. She could buy ten covers with that—or send away and get some secondhand books. She shook her head vigorously. It wasn't a contest anymore. She loved her beautiful pictures and it did not matter that she might not have as many as Elene. It did matter that she had pictures that her father had as a boy.

There were hours every day in which Mandy had nothing to do in school, except listen to what was being taught in other grades. That's what came from having five grades in a room. She learned a lot about a grade before she got to it. Mandy could already get a high mark on a grade Xl history quiz. Today she was bored and sleepy as she watched large splatters of rain hit the windows. A poem she had written drifted into her head like a lullaby:

> *O rain! how tender, sweet*
>
> *your gentle feet.*
>
> *They patter fast,*
>
> *their spell to cast,*
>
> *to lure me to your arms of sleep*
>
> *while you, all night, your vigils keep.*

To keep from falling asleep, she reached inside her desk and took out her covers. A deliciously pleasant feeling enveloped her as she looked at her "old covers". Then suddenly she met the cold, dark eyes of Mr. Rachit.

"Bring that here," he ordered.

She was so startled she didn't move.

"Well!" he stared.

She pushed the chair back unsteadily and, holding her covers tight to her chest, stood up. A kind of mesmerization took over her body as she moved like rubber towards the teacher's desk. She held her precious pictures out hesitantly, careful not to touch Mr. Rachit's clammy hands. As she lifted her eyes to his face, she could see that his cheeks were dabbed with the colour of plumboys. Temper signs! Her heart turned inside her as heavy as a stone.

She watched him walk toward the garbage can. He must be going to throw them in, she thought, and promised herself that she'd get them out later. Then to her horror he began tearing them one by one. She watched little pieces float as useless garbage towards the garbage can. Each rip was like the point of a protractor stabbing her soul. Her loss soaked through her like a poison, weakening her so much that she didn't think she'd make it back to her seat.

She was surprised to find herself moving back to her desk as if nothing had happened. If only he'd strapped her fingers until they burned, and beat her fingertips until they blistered. That's what her teacher in kindergarten had done, after he had warned her that he'd click her if she didn't stop changing her pencil from the right hand to the left. It was easy to survive that. She'd never get over this. There was a tearing at her heart as she thought of the special picture she had been going to frame in Popsicle sticks for Father's Day. Her father would have loved the black and white picture of an old sailing ship rising gallantly on waves that curled around its bow.

The other pupils went on with their work as if nothing had happened. Only Elene looked at her. Then she smiled. Mandy met her eyes as if she didn't care. But inside her, anger was building. It pushed away the shyness that she always felt, and as soon as school was out she walked determinedly towards the garbage can, her skin quivering with rage. She reached down and picked up every jagged piece of picture, wiping away the dust and lead particles that had come from the emptied pencil sharpener above the can.

Mandy took the pictures home and taped them back together with her father's black electrical tape. Her mother didn't know why she was wasting her father's expensive tape, and she didn't say anything against her teacher. After all, it was not as if she had come home with marks from a strap.

The hurt that had come with having her pictures destroyed made Mandy bold enough to take them back to school. She held them in front of her when she was supposed to be working and looked the teacher in the eye, as if daring him to take them away—again. All morning she did this, and each time, he turned away. When she went home she threw the torn pictures under the damper of the kitchen stove.

She might never have put the loss of her pictures at the back of her mind if she hadn't heard Elene talking to Dora about an artist her mother knew. He was coming to Maley's Cove during the summer holidays to open an art studio. He would be using the abandoned one-room schoolhouse that had been replaced by the three-room school that Mr. Rachit taught in.

Mandy could just imagine how wonderful it would be to learn to paint like Norman Rockwell and those other artists who'd painted some of the pictures reproduced on scribbler covers. She wouldn't bother to paint the cove. It was already hanging in most people's windows, with its moods and its colours changing with the weather. She'd paint people in settings that told a story at a glance.

She lost the teacher's voice more than once during those last few days of school. Then finally it was over and she fell headlong into summer.

One morning, a stranger knocked on the door and asked if he could borrow some detergent.

"Detergent?" Mandy hoped she pronounced it right. He nodded and she went to ask her mother.

Her mother frowned. She'd never heard the word, "'Tis probably some new word for some old one," she said.

Mandy wasn't about to ask what it was. She went back and told the stranger her mother didn't have any detergent.

He ran long fingers through thick, curly hair. "That's too bad," he said, "I was hoping to wash my car in the pond."

"Oh!" Mandy exclaimed, "just a minute, I'll get you some Sunlight suds."

A generous smile spread across his soft, smooth face when she passed him a cup of suds. He turned to go, then stopped. "What a great view!" he exclaimed, nodding towards the pond.

Mandy followed his look to the boiling, dark water banging itself against the rim. "That!" she answered scornfully. She stopped—suddenly realizing that scenes don't have to be peaceful for some people to like them. Her voice softened. "You should see it when there's not one ripple. Then you don't have to look up to see the clouds and the sky. You can see them painted on the water."

"Speaking of painting," he said, "I'm starting an art class in Maley's Cove this summer."

"You're an artist," she stammered.

"Yes, my name is Tom Adrian, and you can come for free. The government has given me a grant that will take care of all supplies. We'll be starting on Monday."

Mandy's mother wasn't pleased. Her voice was blunt. "Painting pictures is a lot of foolishness. If you want a picture of the cove or anything else, you can take one in the blink of an eye with a camera, and the camera gives the right likeness—not that blurred mess some artists turn out—like the ones in your father's *Time* magazine." She turned away from Mandy to beat a mat against the fence.

"But I want to paint—look at that mat. You spent all winter breaking your back and stabbing your fingers just to hook the picture of the Newfoundland dog that you'd marked out on burlap. When someone wanted to pay you good for it, you wouldn't hear of it."

Her mother ran a stubby hand across her creased forehead and smiled sheepishly. "I guess we all got our interests," she admitted.

The air was warm and fresh-smelling as Mandy skipped over the hills to her first day at the new studio. She stopped at a muddy pothole and licked out her tongue at the milky image of herself upside down. She passed Uncle George's door, then froze when she saw him lift his axe and bring it down on a heavy, scrolled frame trimmed with gold.

He reached for another, and Mandy cried out, "Don't, Uncle George." She ran up to him. "That's a picture frame."

He stopped, spat on the ground, then wiped his sleeve across his sweaty brow. His eyes looked like two shiny dark holes in a spy glass. "That it is, Girl—but this is an old one—and a good, dry piece of firewood. I'm clearing out the whole lot." He pointed to a pile of split frames. There's new frames to be had these days, ones not as heavy and not as likely to harbour dirt." He lifted his axe and let it come down hard. The frame split and Uncle George tore the pieces apart and flung them on top of the pile.

Mandy walked slowly past him wondering why people destroy things that other people put their hearts into making. Suddenly she began to walk fast—then faster, reaching the landing beside her grandmother's house out of breath. When she saw Mandy, her grandmother laid down the basket of clothes she'd just come out to hang on the line.

"What is it, Child? You look troubled."

"Scared, Gran—I'm scared—I'm afraid to ask if you've been attic cleaning." The words tumbled out, "—and you've cleared out all the old stuff—or given it to some antique dealer rummaging around for old things like picture frames."

"Not at all!" Her grandmother waved away the very thought. "There's lots of frames in the attic—though the pictures are nothing to be excited about. They're just copies of famous paintings. You can have the frames. They're a little fancy for someone like me—too old to see the dust in the crannies."

Mandy followed her grandmother into the house, and climbed the ladder to the attic. I won't tear the pictures, she thought. I'll just cover them with my paintings.

"Perhaps," her grandmother mused, "you're like your great-uncle James. He loved to paint. Look at that picture he did of the *Titanic*. He did it with charcoal, and matches he split with his teeth to make brushes—just come of age, he had, when he went down on the *Southern Cross* and lost his chance for fame."

Mandy got the surprise of her life when she tried to paint. All the beautiful pictures she could see in her head wouldn't come out—no matter how hard she tried. She gave up trying to paint people. It was easier to take colours and let her hands run free with them. When she asked Mr. Adrian if he liked her paintings, he looked a little troubled. "Learning to paint pictures takes time," he said cautiously. Then he smiled gently, "Your colours are passionate and your painting versatile—and there is a name.... Let's just call what you do abstract art."

She was itching to ask what he meant—but she didn't want to show her ignorance. She'd look it up in the dictionary.

"That's not art!" was Elene's scornful observation. "Art looks like real things. You should not have to guess what's there."

Mandy responded haughtily, "Mr. Adrian likes my paintings. He says they are versatile and abstract. I looked the meanings up—and they mean that people can hang my paintings any way they like: upside down, or right side up, and see what they want in them."

Elene lifted her chin and raised an eyebrow, "Go ahead and make a mess, you—"

Mandy interrupted her by pulling a bunch of bills from her pocket. "I got enough money for my paintings to buy some Salle Lee books the next time I go to St. John's."

"I know why your paintings were bought," Elene taunted her. "Tourists from "The States" can't help themselves from buying anything as old as the hills, like the frames on your paintings. I heard one woman say that the frame was worth over a hundred dollars and she had to pay only five dollars."

Mandy stiffened. "I don't believe you," she cried. Her eyes widened in horror as Elene pulled a picture from behind her back. The frame was gone.

"My painting," she whispered, "where did you get it?"

Elene ignored the question and tried to imitate the tourist whose wife hadn't wanted to buy the painting, "That, my dear, may be blobs of paint, but the frame is a work of art."

Elene struck up laughing, "Don't worry, you can paint me."

Mandy's eyes flashed, "I would—I'd paint you if only I had a paint brush and some tar; then I'd hang you in the frame of that old outhouse that leans over Kennedy's Cliff."

Tears welled around her eyes and *peased* out. They fell without a sound from her as she ran home and into her bedroom closet. She sat beside her mother's box of mat rags with her head bowed to her knees. The picture of herself as an artist had been torn. She cried until she felt she couldn't squeeze another drop of water out of her, but as she left the closet she didn't feel lighter. It was as if her feelings were all tangled inside the concrete weight her head had become. She'd talk to Gran. She'd know what to do.

Her grandmother looked at the painting she brought to her, turning it upside down, and sideways before speaking.

"It's called abstract art," Mandy explained uncertainly. "You let your hands do whatever they want—and anything can happen. It makes me feel free. Perhaps other artists decide what they are going to be abstract about. I don't." She felt defiant.

"Is this how you see the world?" Her grandmother's strong, dark eyebrows knit together. She looked concerned.

"No—no—" Mandy cried. "I see it clear—and beautiful—too wonderful for words, but my hands are blind. They can't see the pictures I take inside my head. Now people have gone and done this to me—made me think I can't paint."

Mandy's grandmother lifted her granddaughter's chin and spoke bluntly, "No one has done anything to you. You've learned the truth as other people see it. If you really want to paint, perhaps, with a lot of hard work, you can. You've got long fingers for something. Still, pencils and scribblers might be the clear thing for you. The right person can paint the world wonderfully with words, and even show pictures of people's feelings, just as X-rays show broken bones. Now look at your eyes all puffed. I can tell they've been leaking like a fish flake in the rain."

"I'll never cry again, Gran," she promised gamely, then thinking herself a little rash, added, "not over torn pictures, and paintings torn from frames, anyway. I'll just go back to the studio and take my pictures home."

As she walked up the wooden steps and lifted the latch on the door, she heard Mr. Rachit's voice. She walked in just in time to hear him say, "I'll pay ten dollars for that painting." He turned to look at Mandy, then he added in a quiet voice—"without the frame."

He broke the heavy silence between them. "The ten dollars will help you buy more scribblers, and you can start a new cover collection."

"I mean to buy scribblers," she answered in a determined voice, lifting her head so her eyes were level with his, "but I won't be saving covers."

"Oh?" he looked surprised.

"I'm going to paint pictures with words. People will read my stories and see themselves."

He looked at her, as if he wanted to say something else. She didn't wait for any more of his words. She flung her hair back as if to sweep away all her disappointments, then she went outside.

The sea was quiet, its gentle heaving fracturing the sunlight. Flecks of gold danced across its green waters. It's just too beautiful for words, she thought.

Well, almost....

Suddenly "the witch" jumped in front of her. She had never been so close and Mandy drew back, startled.

Aunt Jerusha squinted at her, her eyelids easing down over her eyes in doughy folds. "What would you give me if I gave you the $100,000 coupon—your soul?"

THE WITCH

Mandy lay in bed, her eyes half-opened. Squeezed between her eyelids was the shadowy image of a girl on a bicycle riding down the smooth, sun-baked road outside her house. The image fluttered with her eyelashes as if the girl on the bicycle was actually moving. She tried to keep her lashes still so she could hold the picture that was in one block of her new patchwork quilt. If she thought about it long enough, the little girl became her on a bicycle, listening to the sound of her wheels swishing on the pasty-coloured road. The sun shining in through the white, lacy curtains made a pattern on the wall that danced, pulling her eyes open. Instantly, she saw that the little girl in her patchwork quilt wasn't riding a bicycle. It was all in her imagination, just like her dream to have a bicycle. And if the witch could do it she'd be walking the rest of her life.

Mandy knew that Aunt Jerusha was not the kind of witch who shows up in Halloween pictures: one with a long nose, a shovel jaw, spider-leg eyelashes and hair like a birch broom in fits. She dressed normally in a black dress with a white, starched flour bag apron. But she did have a birch broom, one that she let fly at dogs that squirted against her door. Everyone knew that if she came out the front door she went in the back way, but that was just common sense for a lot of people who didn't want bad luck—people who believed there was a lot of it around.

Witches were burned in Bible times—and once in Maley's Cove. No one knew if Aunt Jerusha's grandmother, Endora, who lived on the cliffs above the wharf, was really a witch, or if she had placed a hex on the two fishermen who drowned that summer. It was Bonfire Night when her house was caught on fire as she slept. Someone said there was a blaze of light and it was as if Armageddon had come. People saw a flaming, flying figure in the air before it splashed into the sea. Later when Endora was found, they said she was as black as the black man: their name for the devil.

The fishing season was grim the next year, and times were poor for awhile. People who would have nothing to do with witch burning, grumbled that those who did were the cause of hungry times. "They were asking for trouble burning her," someone warned, "witches can do more harm as a spirit no longer tied down by a body."

Aunt Mary, Endora's daughter, wasn't in the house at the time. She grew up an orphan, and married a fisherman who didn't believe in witches. Still, some people wondered if their daughter, Jerusha, didn't carry witch blood in her veins.

Aunt Jerusha seemed sensible enough, and she wasn't skinny, or like something stretched as witches were portrayed. Her legs, like stumps, were in flesh-toned stockings, and there were no bumps where ankles usually are, just two legs standing in sturdy shoes. After a summer in the sun her face looked like a pudding bag turned inside out with molasses and raisin imprints still there, and when she screwed it up, it looked like a brown crinkled bag. She wore a black hairnet on her hair and a black dress with a little bit of white lace around her neck, reminding Mandy of a painting of Queen Victoria that was on her grandmother's wall.

Aunt Jerusha lived in a red house with a black trim and matching black door. Mandy had to pass it every time she went to Aunt Sara's shop for her mother's groceries. As soon as she came near it, the power of fear took hold of her. What if "the witch" was reading her mind and spelling ways to stop her from having a bicycle?

Jeff had a bicycle. Her brother bought it with the money he made from fishing with his father. Mandy wanted a bicycle with a basket to carry groceries and books. It was hard walking to and from the shop and school, especially when the wind whipped her face, its steel digging into her knees and face like nails. Her hands got galled carrying packages of groceries, and sometimes the string burst off the brown wrapping paper and she had to put things into her pocket. Having a bicycle would make her feel she was flying home, compared to walking, but she had no way of making enough money to buy one. Only her mother paid her for baby-sitting, which wasn't much. None of the other mothers did. They acted as if "the idle hands" of young girls were just waiting for the privilege of caring for their children while they had a night out.

When she found the piece of metal in a tin of corned beef she opened one day, she thought it was her chance to make some money. Her mother was down in the fishing stage *bulking* fish at the time. Without telling anyone anything about it, Mandy wrote the company, promising not to say anything about what she'd found if she received a hundred dollars. She was pretty sure she'd get it. Her mother would be shocked and surprised that she tried to extort money, but there would be nothing she could do about it. The money would be rightfully hers. The company could be thankful she hadn't cut her tongue in half on the sharp metal. Worse, someone could have choked on it. A hundred dollars was only a small compensation for all the things that could have happened. Beside, she had let the corned beef company off with enough. She hadn't asked for anything when she'd found a spider in a slice of meat she was eating. She tried not to think that a few of its legs were missing, consoling herself with the thought that she wasn't sure how many legs the spider had. What came in the mail was heavier than money—even if it had all been in loose change. What she saw, to her dismay, were two tins of corned beef.

Aunt Jerusha must have read her mind, and put a spell on the beef company. Mandy felt her watching from her door every day she went by. If she couldn't see her, she always imagined a dark shadow in the window.

One day when her mother sent her to the store for an Ogilvie cake mix, she saw on the back of the box an offer of a bicycle to anyone who saved coupons with numbers that added to one million dollars. A different denomination could be found in each box. Here was her chance. She was determined to save that many coupons even if everyone got sick of eating cake. "Let them eat cake," the words danced in her head from somewhere in her history book. Marie Antoinette had said them. Never mind! There was a variety of cake mixes: chocolate, lemon, cherry, and even lime, and Mandy's mother was willing to buy them. Her father would probably have to grunt his way through the store-bought mix after having them a time too many instead of more traditional desserts like blueberry and partridgeberry puddings.

She almost lost hope when she thought of Jeff drawing a perfect likeness of Robin Hood for a Robin Hood contest. He had been very careful with measurements and it was strange that he didn't win. But the Ogilvie contest was a sure thing! It had to be!

Thinking of her bicycle made her feet fly every time she went up the road now, and the heat of summer days wasn't so bad. At home, she went around the kitchen singing while she was doing the dishes. Her mother looked at her as if she was speaking in tongues when she began singing a song in Swahili that a missionary from Kenya had taught Mandy's Sunday School class.

Sometimes there were boxes with no coupons in them. Some mean person at the factory where the cake mixes were packed probably left them out, thought Mandy in frustration. The expiry date was getting close and she still needed a lot of coupons. She imagined getting more than the $100 coupon that was in many of them, and once she dreamed she had a $1,000,000 coupon. Her mother was beginning to worry. The crease in her forehead seemed to deepen as the time drew near and coupons with a low value were all they were getting. "So close," she murmured, "and yet so far." It wasn't enough to get almost a million in coupons. Just one short and her dream of getting the bicycle was over.

Aunt Jerusha saw her one day when it was raining. She watched as the cake mixes, along with a tin of Carnation milk, fell through the wet wrapping paper to the ground into a puddle of water. "What's all the cake mixes for, me maid; t'ree is it, yer got dere?"

Mandy's insides trembled when the old woman spoke to her, but not wanting to make her angry, she explained what she was doing, figuring that the woman had X-ray eyes and knew anyway. Her coat was wet and muddy from where she'd picked up the cake mixes and milk, and the old woman didn't offer her any wrapping paper to hold her groceries in. She'd just have to tie the shopkeeper's twine around everything and hold it in her arms. Carrying parcels in her arms was so tiring. She really wished she had a bicycle.

"You won't get it, you won't," Aunt Jerusha crackled. "Yer legs were made for walking, Girl, and dat's what you'll do, mark my words."

Mandy's heart sank, but her will rose in her strong and angry. No old witch was going to keep her from getting a bicycle that she'd worked hard for, a bicycle she had even prayed for. She knew that God didn't take up for witches. In the Bible they were burned, except for Moses who made rods turn into snakes, and Elisha who put curses on people, but they weren't called real witches. She read a story about a witch once, just before she went to sleep. Afterwards she got afraid that someone was coming into her room after her. When she finally fell asleep she dreamed she was dead and laid out cold on the floor in a very cold room. The old woman was talking to her while everyone left the room. She couldn't run any longer, and the woman was coming closer, getting ready to live in Mandy's body. She had been glad when morning took away the darkness and her dreams. Now she wanted to be past Aunt Jerusha for fear she'd take over her body.

Jeff laughed at the idea of anyone being afraid of Aunt Jerusha. He mimicked her as he forked mashed potato into his mouth. "I'll have yers shot." That's what she'd shouted at the boys for stealing her apples. "You'll wake up der marrow marning ter find yerselves dead." It had been so funny when she had started shooting blanks and Jeff had run down the road shouting that he had been shot. Jeff's friend, Mark, said that his mother had told him stories about "the witch." She had once drawn a picture of a woman she didn't like on her wall, and then put a nickel in a gun and shot her in the arm. The next day the woman had a broken arm. She was lucky it was only her arm the witch had shot at.

Mandy's mother shook off the stories, calling Aunt Jerusha an old woman not right in the head, but not evil. "Lies," she said pointedly, "are worse and more harmful than she is to the community."

Mandy was afraid to mention to her mother her fear that Aunt Jerusha had put a spell on her. She'd think she had demons like Mary Magdalene, and they'd have to be cast out of her. If she could only find a four-leaf clover like the one Aunt Callie had in her Bible. The leaves formed a cross that was used to ward off evil.

It had good fortune against witches. Her mother wanted to know what Mandy was doing when she saw her kneeling on the ground out by the clothes line with her hands in the grass.

"Oh Mom," she whispered, "Aunt Jerusha has gone and put a spell on me, and I can't find a lucky clover." Then she remembered something. She stood up hopefully, "I'm going to make a charm against her." She pulled a piece of her father's twine from a pocket in her gored skirt. A cent she'd found in an egg between the velum and the yolk, when she was at Aunt Callie's, was surely lucky. She'd bore a hole through the cent and put the string through it, then she'd tie the charm around her neck.

"Stop your nonsense," said her mother. "Jesus owns you. Aunt Jerusha is just a harmless Church of England woman who can't put spells on anyone. She's probably just having fun with you. Your face opens up to every kind of emotion. She sees your fear. Maybe that's the only excitement she gets. She doesn't know what it's doing to you."

The final day for the bicycle contest came and Mandy still needed a $100,000 coupon. She wasn't getting a bicycle the easy way after all. After she pushed the money for the cake mix towards Aunt Sara, she hurried outside and leaned against the porch on Aunt Sara's shop. Her trembling fingers tore off the bottom flap of the box. She was so full of hope she could hardly breathe. The coupon was always on the bottom. She pulled it from the dusty flour mix, and shook the powder off. Her hope dropped. It was a $50,000 coupon.

She backed into the door, turning the knob. She'd buy another cake mix. She still had thirty-nine cents. Her mother wouldn't care about her spending the money if she got the coupon they needed. Aunt Sara looked at her inquisitively over her wire rim glasses. "You and Aunt Jerusha, I declare, I don't know which is worse for these cake mixes. Your mother's got a crowd of youngsters, but Aunt Jerusha—she must feed the crows."

That's odd, thought Mandy. She never ever smelled any aroma as wonderful as a cake baking any time she passed the witch's door. Then a thought grabbed her and squeezed her heart. That old woman was trying to get all the good coupons so Mandy wouldn't get them.

Aunt Sara turned to the shelf behind her. "Um, I guess they're all gone."

Mandy wanted to cry—and she would have if she hadn't been too old for such childishness. Instead, she pushed the money back into her pocket and started home. She didn't even look towards the witch's house as she passed.

Suddenly "the witch" jumped in front of her. She had never been so close, and Mandy drew back startled.

Aunt Jerusha squinted at her, her eyelids easing down over her eyes in doughy folds. "What would you give me if I gave you the $100,000 coupon—your soul?"

"Nothing—I can't," Mandy protested. "I don't own my soul."

"Then who in the world owns it?" she asked in a crackling voice.

"No one in the world," Mandy explained. "God owns it—so I can't give it away. Besides, you don't have a coupon."

The old woman pushed her hand down a deep pocket in her white apron and pulled up what looked like several Ogilvie coupons, one with the number on it that Mandy had dreamed of and imagined having so many times. She shifted her parcel to one arm and reached for the coupon.

The old woman laughed, a glint in her eyes. She held it up. "See—a $100,000 coupon—it's real. It can be yours—or paper for my fire."

Without thinking about anything, Mandy suddenly reached out and grabbed the coupon. The old woman looked at her in surprise, and then she laughed. "There's a spell on it. It won't do you any good."

Mandy didn't speak. She was so shocked at what she'd done, and then just as shocked that she had the coupon in her hand, that she started walking as fast as she could go. She was home before she knew it.

She caught her mother's smiling face in the window as she bounced up the steps and rushed inside. Her voice sounded relieved, "I could see by the way you skipped down the road you had your coupon."

"And we won't have to eat any more of these stupid cake mixes," Jeff said, looking into the mirror on the wall and pushing Brylcreem through his dark, curly hair.

After the coupons were counted and checked, then mailed, Mandy spent hours imagining the bicycle was standing there waiting for her like a pet horse. There were days that she hurried to the post office hoping to receive the notification that her bicycle had been shipped from Montreal. She hoped her coupons—and Aunt Jerusha's—wouldn't be like Jeff's Robin Hood drawing, and disappear without a trace, or the hundred dollars she had thought was as good as in her hand.

As the weeks passed, her mother cautioned her about getting her hopes up. There could be all kinds of reasons why she hadn't heard. Mandy began to imagine a few. What if the coupon she got from Aunt Jerusha wasn't real? What if she really was a witch and had conjured the coupon and the judges had been able to tell it wasn't real?

Then one day her father came home with the bicycle in the wooden box of his old grey truck. He walked into the house, smoothed the thinning hair across his head with his fingers and smiled at her, "Well, Mandy, you got your wish."

She couldn't believe it. The smell of rubber was heady in her nostrils, something like the rubber on her new pencils the first day of school. The chrome, with its sky blue ribbon of colour, was more beautiful than anything she'd ever seen on a bicycle.

Mandy sat down on the soft, black seat, letting the 14 blue and pink gores of her skirt flare out around her like a Chinese hat. She knew how to ride her bicycle. She'd sneaked Jeff's when he was out in the boat. The first time she tried to ride it, she fell off, banging herself hard on the cross bar. She might not have an amen or a hymen (whatever it was called) after that. She wondered why boys had bars across their bicycles and girls didn't. They were as likely to do themselves damage as girls were. She'd remembered some guys sniggering that when young Ebbie got on his new bicycle, he went over a rock and banged himself so hard.... She wouldn't dare say the words...., nor would she have the face to ask her mother if such accidents happened. She would call that kind of talk dirty blaighard.

Lucky for her, Jeff wouldn't want to ride a girl's bicycle. His bicycle looked shabby against hers. He was always patching one of the tubes, after running over pieces of glass or nails. She was going to be very careful where her bicycle went.

"I daresay there'll be a problem getting parts for this one," her father surmised, "since it's not a Raleigh."

"Parts! They're all there," she exclaimed.

"Now, they are," her father's forehead crinkled, "but you never know what can happen."

Nothing could! The bicycle was like a million dollars—a million paper coupons magically turned to dollars and then into a bicycle.

She was so happy she almost cried, but she stopped herself. Crying was a silly thing to do, and Jeff would laugh and go tell everyone.

Aunt Jerusha's face came to her. It was going to be a pleasure to ride past her—fast and unafraid.

She wasn't scared riding down around the pond on a pitch black night. What could harm her? One reason she didn't feel afraid, despite the tales of ghosts and pirates' gold, and her brother's story of masked men hiding in the hills with sawed-off shotguns, was that she'd be off in her own mind, wrapped up in her imagination away from the darkness that came down around her as thick as a shroud. Her thoughts and imagination carried her along, away from her fears, away from her actual, physical self. This time she heard a noise that drew her back inside the dark—back into herself. But she was on her bicycle. She could pedal fast even if she didn't have a light on her bicycle yet. Just as she picked up speed, her front tire hit something dark and solid. She was sure hair—not her own—brushed her face as she hit the ground. She jumped up, remembering afterwards as she ran with her heart beating inside her head like horses' hoofs that there had been a noise. It must have been Aunt Jerusha, dressed as a black ghost, trying to frighten her out of her skin.

She turned the foot of the pond and ran towards the light over the door of her house. She got nearer panting and gasping for breath. To her horror, the house was surrounded by horses. She stopped and tried not to breathe too loudly for fear of making them gallop towards her. A saucy-looking one stood close to the path leading to the house, and kept looking at her. She tried to walk softly, wanting desperately to call to her father to come and drive them away. Every time she put a foot forward she heard a snort, then a trot. She finally got enough nerve to rush towards the door, turn the knob and stumble inside the house.

Her mother, peeling the vegetables for Sunday dinner, as she did every Saturday night, turned and asked, "What did you see— a ghost?"

She didn't answer. Suddenly she remembered that her bicycle was lying on the road. In that instant she was outside, uncaring about the horses or ghosts, running as fast as she could up the road. She had to get to her bicycle before the vehicle coming down the road hit it. Headlights showed it was getting nearer. Suddenly she saw the back wheel of her bicycle rise into the air as if it were alive.

The handle bars were twisted out of shape, and the tires' rims were bent. The bicycle seemed to drag away from the pull of her hands. It seemed to take her a long time to get it home. She passed the horses without even looking at them. Whatever they could do to her would not be as bad as what had happened to her bicycle. It was her own fault for snatching the coupon. Aunt Jerusha would laugh when she saw Mandy without her bicycle.

If her bicycle was different before, it was really different now. Unspeakable sadness clouded her mind. She felt all filled up with fog, heavy with tears that finally burst from her eyes and ran down her face. Her eyes raked her father's face as she pleaded, "Fix it Dad."

"Fix it," he grunted. "No man on earth can fix this thing." He tried to straighten the handle bars, but he couldn't stop them from curving up into the air like the letter M, and the tires were wobbly even though he tried to straighten them. The mudguard made a soft swishing noise against the tire of the front wheel. The brakes weren't working well either. She'd probably wear out the taps of her shoes trying to brake on the hills.

She decided to take the bicycle and ride it past Aunt Jerusha's, trying not to notice that because of the twisted handle bars her behind was stuck in the air. As she came close to the downhill slope by Aunt Jerusha's, she smiled at her upside down image in a calm, muddy pool of water. She saw the next puddle too late. She bounced right through it, sending a splash of gritty mud into her face.

She heard a sound like a cackle, and she wanted to cry, but she didn't. She went on down the hill with her behind stuck in the air, putting her foot down whenever she needed to brake. She wasn't afraid of spells. If her dreams for a bicycle could come true once, the possibilities for the future were endless, and she wasn't going to let some old woman, who only thought she was a witch, tell her future—or even a little part of it.

"Stretch up!" she called. "Stretch!" She closed her eyes. The cold wind was whipping her dress around her legs. What if he falls on me! she thought. We'll go banging down the cliff like old cans. No, not cans—pulpy oranges—and our skin will tear off, and we'll drown in the ocean.

A MARE'S NEST

Mandy woke to chirping sounds filling the spring air. Robins, perched on wires outside her house, bathed her in their music. Birds don't really sing, she thought, they play music. Her eyes opened to dust dancing on a shaft of sunlight that streamed in through her bedroom window, shining the dull red canvas. She stretched her arms into the air, awaking her whole body to this world of enveloping peace.

Then she remembered what had happened, and suddenly the high-pitched sounds of birds seemed to peck a hole in her heart until she felt it was going to collapse inside her. It became a muffled beating inside her head. She had loved Scruples, even though she hadn't been the most beautiful cat in the world. Jeff said she looked like something left out in the rain to rust after stirring a can of black paint with her tail and ears.

The cat had been missing for five days before she came home for the last time. Mandy was lying on the daybed reading, and pretending she didn't care that the cat hadn't come back, but behind the book tears were streaming down her face. She was

145

almost to the end of a Salle Lee Bell book when she heard a scratch on the storm door. Her book was thrown to one side and she jumped up to open the door. Her father got to it first, and the cat limped across the floor, lay down and went still. As Mandy rushed to grab her up, her mother cried, "Don't touch her!" She saw then that her cat's side was ripped open, and infected. Her father said that it looked like the work of boys with pocket knives and pellet guns.

Mandy knew that some of the boys in the cove were capable of anything. She'd seen Al and Joe pelt an old tire with rocks and shoot BB pellets to kill baby birds newly born in a nest their mothers had made in the tire. Sometimes the boys fed the birds to stray cats, then smashed the skulls of the cats. She never thought her cat would be a target.

"Now the birds are safe," her brother said, grinning.

"Only from the cat," she answered angrily. She'd chided her pet when she found feathers from a robin outside the door. She hadn't been able to convince her cat that birds are not cat food.

If her cat was going to have to put up with vicious boys, she was glad she had died. She wondered how much distance there was between cat Heaven and where humans go. Her mother didn't believe there was a cat Heaven. She said cats didn't have souls. She couldn't explain how she knew that. No one had ever seen a soul, not even in humans. "There's no comfort in believing in something false," chided her mother, "so I don't want you thinking that your cat's alive somewhere."

A lonely feeling came with her mother's words. After all, no one had ever seen paintings of Jesus with a cat on his knee—a lamb, perhaps, but not a cat.

The next day she and Taylor, her younger brother, found a lamb. They were coming down the road and had just turned the corner at the foot of the pond when they saw a pure white lamb that had strayed from its mother. It was bleating and Taylor ran after it. "Don't!" Mandy called angrily, "you'll frighten it." The little lamb tottered on the edge of the bank and fell into the cold water. Mandy tried to pull it out, but it was all sogged, and heavy. She ran home panting, "He fell in—he fell in," to her mother who was stirring a pot of soup on the stove. Her face went white as she turned frightened eyes towards Mandy. She put the spoon down, and lifted her hand to the waves of black hair across her forehead.

"A lamb fell in," Mandy said. "Where's Dad?"

"He's in the basement—and don't frighten me like that again," her mother warned, the colour creeping back into her face.

Mandy had burst into tears by the time she reached her father, but she managed to tell him that there was a lamb in the pond and she couldn't get it out. Her father straightened up from the runners he was putting on a new wooden *slide* and wiped his hands on his trousers. Without a word, he hurried out the door. Mandy followed him. Taylor was holding on to the bleating lamb, its wool lying in damp curls all over its body. It was heavy for her father, like a soaked rug, but he got it into the house and by the stove where Mandy talked to it, never thinking it would die. When it shuddered and dropped its head to one side, she cried. No one else seemed to mind. People killed lambs every day to eat. But Mandy had seen in its face pain and fear, the same things people show in their faces.

In the cliffs of Backhill Rocks, Mandy climbed ledges looking for birds' nests. She always brought worms. There were times she had to be careful when she walked through bushes. Sometimes there was a sudden rushing sound as she walked along the path. She had almost stepped on sparrows' nests and baby birds hopping into the bushes. But often, when she carefully parted the sharp twigs of bushes she made delightful discoveries. There was the narrow, deep sparrows' nest with tiny brown eggs, or little birds. Some birds were covered in down; then feathers began to come out of their bodies and cover them. When they opened their mouths she sometimes spat in them in case they were thirsty. After all, she reasoned, the mother didn't feed her babies milk and she couldn't carry water in her claws.

Two days after her cat died Mandy hurried to a robin's nest to see if the four babies there were getting feathers. Everything was quiet. There was no sign of Mother Robin perched upon the grey cliffs, ready to fly from cliff to cliff and scold mercilessly. There was not even a chirp from the baby birds. Everything was deathly still. She was always surprised how fast the birds grew. One week they'd be scrawny shivering creatures with beaks wide open, and white down above their small beady eyes. Days later, it seemed, there were four birds tightly packed in the nest, feathered and ready to fly. She drew near and walked carefully up the side of the cliff. Bending towards the jutted shelf of cliff she looked in. The nest was filled with cut up birds. She screamed and her voice echoed back to her from across the hills. She ran home crying, "Al tore the legs off the birds, and then he cut the birds to pieces."

"The devil gets into young boys," said her mother, wiping her hands in her apron after washing cod tongues. She laid them in the frying pan where they lifted and hissed in the heat as if in protest. (Mandy never felt sorry for fish.)

"Man shall have dominion over creeping and flying things," said her father, lifting his head from a *Time* magazine. "But wasting the lives of birds for a game, that's something else—that is." The lines along the side of his face deepened as if he was annoyed. After that her father was on the watch for birds' nests, and once when he found one by the side of the road he placed a big rock by it.

Mandy had always watched for robins searching for material. Their nests were usually made of twigs, moss, and dead leaves and lined with human or animal hair, their bowl-like structure shaped by the mother's breast. It was hard to protect sparrows' nests. Sometimes she would almost step on one, just as a sparrow rushed up close to her face. Her surprise was suddenly replaced with the joy of discovery. She would place a large rock a little way in front of the nest. Sometimes she was disheartened to find that the eggs were left cold.

One day Mandy took the path up the lane by the pond, and the bend in the side of the cliff, hoping to see a *beachy bird's* nest. Small sandy-coloured rocks, sharp with myriad shapes, had broken from the cliffs of hills and dropped onto the path she took to school and the grocery store. As she hurried along, Al and Joe started throwing rocks at her. They rained against her legs, sharper than hailstones on her face, almost as bruising as the shot from the BB gun that had hit her breast bone once before. She felt angry, but then she thought of Heaven. Maybe she'd get a crown with an extra stone, shimmering like her Aunt Callie's jewellery.

She never thought of telling her father what the boys were doing to her. He was a man who kept the peace by ignoring incidents, even when other people would have thought of them as crimes. When a basement window was broken by a stray rock or a ball, he would replace it himself, sputtering about responsibility only if it happened more than once in a summer. Even then he admitted that a house by an unpaved road was an easy target for stray objects.

Al and Joe scoffed at the idea of birds and animals having souls, feelings and rights. One day, on her way home from the store, Mandy faced them. "I'd like to see you build a bird's nest," she taunted, "weaving it from twigs without glue or tools. You can tear nests up, but you can't build them. Birds can fly and you can't. And they have feelings—like people." They laughed at her, then bent to pick up a rock.

Mandy ran as fast as she could towards the back of the truck that had stopped for her. Ignoring the swirls of dust coming at her from the tires, she ran panting, her mouth open. Her hands were galled from the line that tied the parcels of groceries she'd been holding. She quickly dropped them on the floor of the truck and jumped up onto the tail board. Suddenly something hit her throat.

She swallowed it before she could stop herself. I've swallowed a fly, she thought grimacing. It was probably alive and eating her stomach! As soon as the truck braked by her door she jumped down and ran towards the house. She swung open the door and gasped, "Water, I have to drown a fly." She grabbed a glass off the table and gulped.

She began sputtering, her eyes watering. "Vinegar!" she gasped, "Oh no!"

"A marinated fly, that's funny," an amused voice spoke.

She turned and saw her older cousin from Gander stretched out on the daybed. Melva looked at Mandy as if it was she who was distasteful—not the fly she'd swallowed. Melva's black hair came to her shoulders so flat and straight, it looked as if she'd ironed it down. The ends were teased like a bird's nest and flipped up. She called it the movie star vogue.

Melva had always acted as if she was better than her cousin. When Mandy was little and had gone to Gander with her mother for a visit, Melva had pinched her black and blue and threatened more black and blue spots if Mandy told that she stole her candy-red purse. Whenever Melva showed up with her mother for a couple of hours she complained about the gravelled roads they had to drive over. She boasted about Gander being a town with paved streets instead of potholed roads. Mandy retorted that a street was just a road wearing makeup. She was glad Melva didn't stay long. She was into everyone's business, even asking Jeff if Mandy had a boyfriend. With cousins like her around, she had as much privacy as a two-hole outhouse.

Melva liked creeping, crawling and other creatures better than people. When Jeff killed a bumble bee in front of her during one visit, she accused him of being a murderer. He had rolled his eyes and told her she was as stun' as a bee in cold weather.

"That's the problem," Melva informed Mandy, "people are always being compared to creatures that are not human. He drinks like a fish, eats like a pig, is as cross as a bear... People insult animals, but they are a cut above humans."

"You should know," said Mandy.

"On the other hand," Melva said slowly, "your freckles do make your face look like a spotted pig."

Her cousin was lucky she didn't stay long enough to get Jeff into the mood for playing tricks on her. Otherwise she might have ended up with crawling creatures in her pocket or a dead muskrat around her neck. Mandy knew what that was like.

But sometimes Mandy and Jeff did things together. After their cousin left he asked Mandy to go picking up starfish and *ose eggs*. Their father's *flat* was pulled up on the sand. Jeff looked across the water. Mandy followed his gaze. She knew what he was up to. "Come on," he called, running towards the boat. "Jump in and I'll push it out."

"But I don't want to," Mandy answered. "It's getting cold and my dress isn't warm."

"Girls," he sniffed, and started pushing out the boat.

Mandy turned reluctantly towards the boat and jumped in. Her behind hit the cold damp seat.

Jeff started running with the boat until it left the ground. The boat lurched before righting itself on the slapping green water. He jumped in and grabbed an oar. "Help me row," he demanded. "We're going to Kelly's Point."

"Kelly's Point," she protested, "that's a long way out."

"Only half a mile," he shrugged, ploughing the water with a sculling oar. Kelly's Point lay in the sea like a humpback whale from a distance; up close, it looked long and smooth, its cliffs steep and foreboding.

When the boat hit the sandy beach, Mandy jumped out and helped Jeff pull it in until it was well grounded.

"Beachy birds!" Mandy shouted as she ran along the beach. "I believe there's beachy birds on the point." She lifted her hand to shield her eyes against the sun and looked towards the top of the cliff. "I can't be sure. It's too far away. Bet there's a nest up there."

"Let's find out," her brother called.

She'd seen only one beachy bird's nest. The white eggs in it were spotted brown. She was almost tempted.

"I can't climb up there," she said wistfully. "It's steep, and Dad said it's a hundred feet high."

"Yes, you can. Just follow me," Jeff ran across the path to the cliffs.

Mandy held back. "You're wearing rubber boots. My shoes might slip."

"Come on," he challenged her.

"Okay," her voice went flat.

Jeff put his feet into the small pockets in the cliff and pulled himself up. Mandy, thinking it wasn't so bad after all, followed him.

They kept climbing. Mandy's toes began to hurt from having them bang into the cliff; her hands were getting cut on the sharp edges.

She was sorry they had started to climb Kelly's Point. There was probably no nest there. They would have seen a bird fly out of the cliff. Perhaps it did when they weren't looking, though they should have heard the flutter, unless it had flown away while they were running across the beach. When they were close to the top, she felt excitement lift her spirit. She was sure she could see a nest in the cliff.

"Why are you stopping?" Mandy asked impatiently. "Hurry!"

Jeff half-turned his head. His voice was strained. "I can't go any further," he said helplessly. "You'll have to climb back down."

"Back down!" Mandy cried. "I could never do that. I wouldn't know where to put my feet." She suddenly became aware of the height. She listened to waves bashing against the cliff almost mesmerizing her. Her toes began to get numb from having to stay in one place, and the sensation of being wrapped in skin as heavy as lead got worse. If she moved, she was sure she would fall down on sharp cliffs, banging and scraping her freckles off—which was the only good thing about falling. Her dreams of falling were coming true. She felt cross about that. Dreams weren't supposed to come true.

There was a tree right at the top. Its root came down into the cliff. "Can't you get hold of the root coming out of the cliff?" she asked in a wobbly voice. "It's fastened to the tree growing on the top."

"That's not much bigger around than a pencil," he said scornfully. "If I grabbed it, I'd be down on your face in a second."

"Uncle Gus held on to a tree root once and caught himself from going over a cliff," Mandy insisted.

"Oh no—that's where the nest is," Mandy gulped, almost choking on a mixture of fear and disappointment. "We can't pull on the root. The nest is wedged between the cliff and the root."

"You—and Melva," her brother shouted down at her, "are sissies! Do you think the bird would try to save you? What do you think nature teaches? The survival of the fittest; that's what! Cats eat birds, birds eat wiggling worms, lions eat flamingos. And humans! Sometimes if a mother thinks she can't handle another baby she gets a doctor to take it out of her before it can be born." He very carefully turned his head to look down at her scornfully. "Even Mother Nature poaches. She might get us yet. And you're concerned about a nest that may have nothing in it."

She was distracted by the soft sad-sounding klioo of a gull, and then other gulls and their wild meeew. She almost laughed when she thought of what her father told her that morning about gulls. When the male is courting the female and she calls klioo, he throws up his food and she eats it.

She wanted to be home now—or in a safe place on the path leading home from the berry hills, her legs healthy-weary in hot rubber boots, after being twisted over knobs in hills. She wanted to be where she could see the green and red tiles on the roof of her house, and be able to run free of *gowity* bushes scratching at her clothes down towards the smell of new cabbage and blueberry pudding. She wanted to be rid of the terrible, strained pain of keeping her feet in one spot.

"I'm not going to die because of a nest," her brother called. "There's always birds to have babies. It's not as if we're saving the last bird. Even then... Oh why talk to you!"

"But you might die anyway." She was testing him. "What if you pull on the root and it comes right out of the cliff?"

Jeff ignored her as he worked, cautiously pulling out shale until he could get his foot in and reach for the root that he'd scorned earlier.

Mandy watched as he stretched his arms up. "I can't reach," he grunted.

"Stretch up," she called, "stretch!" She closed her eyes. The cold wind was whipping her dress around her legs. What if he falls on me! she thought. We'll go banging down the cliff like old cans. No, not cans—pulpy oranges—and our skin will tear off, and we'll drown in the ocean. Where was her guardian angel now? She knew where! Frozen in the picture on her bedroom wall, its big feathery wings barring two children from falling over the edge of a cliff. She wanted to climb upon these white wings and fly home.

"I wish I were a cat," she whimpered, visualizing herself running up the side of the cliff on padded feet.

"You'd eat rats and spiders—and birds, and probably get your brains knocked out with a rock," Jeff said scornfully, grabbing the supple root above his head. The nest moved only slightly and Mandy let out a sigh of relief. Jeff lifted one hand and then the other higher up the root until he was up far enough to get his arms around the base of the tree. He pulled himself in on the grassy top and lay still, his head out of sight.

Mandy tried to stop trembling as she clung to the sharp edges with her fingers, her toes dug into the steep cliff. Looking up she could see the nest. "Are there any eggs in the nest?" she called anxiously. She remembered the time she'd been walking up Apollos' lane in the shadow of its rugged cliffs when she spied a nest among the pockets of clay that also held wild irises and bluebells. She'd climbed up the rocks and taken a warm, blue egg from a robin's nest. She opened her hand to show it to Jeff and it had rolled off onto the ground where its shell split into pieces and a mucky mess seeped out, jellyfish-clear mixed with yellow. She had stood still in horror at the splatter on the ground. Then she burst into tears, and ran home thinking of the bird she'd killed while he was still liquid. She was afraid to hold an egg after that, even when it was all solid inside with a baby bird trying to peck its way out. If it was cracked, she didn't mind picking off some shell and velum to help the bird out.

Her brother slowly raised himself from the top of the cliff, and turned to look at her. His face looked as white as some of the chalky rock in the cliff. Mandy could see how scared he'd been— could tell he still was—by his tight, angry voice. "It's a mare's nest," he answered—a mare's nest. That's what Aunt Callie calls something that's much ado about nothing. It's just a nest." He was leaning over the cliff looking down at her. He said slowly, "I'm going to flatten the root down against the cliff so you can reach it." His face turned red as he carefully bent it, so it wouldn't break. "Here, grab hold." He looked relieved.

"I can't," she whispered. "My hand will slip off it."

Just then a mother robin flew overhead, circling and jawing loudly. Mandy looked up, almost losing her balance. Her feelings froze in her as an awareness she didn't want to acknowledge filled her eyes. Suddenly little pieces of her values begin to fall along with shards of cliff. She tried to reclaim them, but fear of dying filled her, pressing against her heart as if to flatten it. She could barely breathe. Her insides trembled. Any minute a piece of shale would slip loose taking her feet with it. She wished Melva was here instead of her.

Her brother strained to push the root so she could get a grip on it. One hand wrapped around it, the other tried to find a place in the cliff to grip.

Jeff called, "Lift your other hand behind the first now that you got it."

Suddenly Mandy found herself hanging from the cliff by a root. She banged against the cliff, bringing sharp pain to her lip. What if the root broke! She closed her eyes and didn't open them even when something brushed by her face. Her feet tried to find the niche her brother had made. Finally she got a foothold and inched her way up. Jeff caught her arms and jerked hard. She flopped to safety on the soft grass.

She lay there for awhile before opening her eyes. She turned to Jeff who stood grinning at her, "I saved your life!" he exclaimed proudly.

Mandy's eyes narrowed, "First you almost lost it—and for what—so we could look for a mare's nest."

She looked down at her torn dress and stockings stained with blood. Then her eyes widened at the unmistakeable stains of broken bird's egg. It must have been the bird's nest that passed by her. "You lied," she screamed, "—I hate you!" She got up and began running, trying to evade the truth that was filling her. Her ponytail banged against her back like a gentle chastising.

Jeff, calling after her, "Nothing is perfect," reinforced her fears that he was right. Animals and birds are part human, but humans are also part animal when it comes to survival.

She felt insulted, and shamed that he had tried to French kiss her. She couldn't tell anyone about it. Her friends would think she was some sort of——. She couldn't say the four-letter word. Her mind would feel even dirtier than her mouth was.

A PERFECT LOVE

Mandy quickly pushed in the brass button that locked the door and hurried across the cold canvas floor to the small table. In the glow of the soft light she wrote nimbly; her pen flew until her fingers cramped. All her hopes and dreams became visible, given to Stephen. With anticipation she signed her name with Ever Yours, and sealed her letter in a matching rose envelope, part of some stationery she'd saved from last Christmas. As she stood up, she caught her reflection in the mirror above the mantle. She wanted to look like Jacqueline Kennedy, wife of John F. Kennedy, President of the United States. She could almost believe that she did until she looked into a mirror. She'd bought a white mantilla to wear on her hair like Jackie's. She was sure she would find a perfect love like hers some day.

She closed her eyes to better picture Stephen when he bent his blond head towards her, drawing in the scent of her Evening in Paris perfume. She loved the midnight blue bottle with its silver cap. It was her special perfume, better than Ben Hur—which was only twnety-five cents a bottle. Evening in Paris perfume cost more than a dollar and seemed to wrap her in love. A smile rippled across her lips when she thought of Stephen stopping by her aunt's gate in Smith's Point where she'd been visiting. He'd told her he was coming to see her.

Suddenly Mandy's smile disappeared as the room was flooded in light. Her father stood in the doorway. "Give me that paper," he ordered.

"No," she cried, standing up. She backed away and he came towards her. He reached for the letter. She held on, not caring any more that it would be rumpled. The words could be rewritten once the paper was smoothed—if she kept hold of it. Anger strengthened her hand. She held on. Finally her aching fingers let go and her hopes were crumbled in the palm of her father's hand. At that moment Mandy felt something inside her go as crushed as the letter. She tore past her father's large frame, crying in rage. Was there never a part of herself she could keep? Must her thoughts and her feelings be always open to her parents?

She ran around her startled mother, whose hands reached out towards her, and darted through the door into the inky night. She stumbled now and then as her feet touched the soft, mucky earth. She wanted to disappear with herself intact, hold herself in the night, cover herself in it so that she could be herself. She made rash wishes, trembling in her power to even think them, wishing thoughts she knew she wouldn't be happy with once her anger had died. But she would never forgive him—not as long as she lived. Her heart was beating around inside her like the hub of her mother's washing machine on washdays. Tired, she sat on a rock, her hair falling over her face. She wrapped her arms around her knees, put her head down and closed her eyes. The grey shroud of heavy fog slipped over her, seeping into the pores of her skin and shook it until she felt like jelly. Shivers slithered up and down her spine.

If only she'd been able to see Stephen before he left for a job in Corner Brook. He had come up to Maley's Cove the Saturday night before he left. He blew his horn and she'd gone to the window, but she hadn't let him see her. She was babysitting her sister, Elizabeth, and she was afraid her father would come back. Stephen had told her in a letter that he'd come to see her Sunday afternoon, but she was in Sunday school, and then Sunday night, but she was in church. That Monday morning he'd left and she hadn't even had a kiss. She would have liked his kiss—just to know what it felt like.

Her father would be looking for her soon, waiting to call, "You get home as fast as your legs can carry you." It seemed a long time, but then she heard him coming, the swish of berry bushes against his legs. She lay down, her head on a rock, holding herself still. Her father passed by and finally went home. She stayed, too angry to be frightened by all the ghost stories she'd heard, especially the graveyard ones. She was close to the cemetery where her grandfather was buried. She didn't care—no one there could hurt her. She wished she was dead. She caught herself. Why did she say that, or even think the words? They were just words of defiance her mind had spat out against her helplessness. But she wasn't helpless. She was full of life. It surged in her against the injustice inflicted on her. It gave her strength; she'd go home and rewrite the letter.

She opened her eyes, suddenly startled as if a spider had climbed a strand of hair in front of her face. She tried to convince herself it was only her imagination, or—perhaps a twig had touched her cheek.

She was part way down the hill when she heard the sound of breaking twigs. Someone was coming through the bushes. She heard the sound of her father wheezing. He was back looking for her, and then he called her name. She stayed still until he passed her, then she ran down towards the potato bed. Clay, made muddy from rain the night before, caked her shoes, making them heavier and heavier each step she took. She was relieved to jump over the picket fence until she felt a strain and heard a rip. Her beautiful flowered skirt, with its pretty crinoline, was torn. Her mother's hand went towards her as she pushed open the porch door and ran past her into her bedroom. She curled up on her bed crying softly.

Her father opened her bedroom door and came in. "I'm sorry," he whispered, laying the crumpled letter on the pillow beside her. "I didn't mean to hurt you." He went out, pushing the button in the lock and closing the door behind him. She got up and opened the door going into the hall. Her hand reached towards the column her father had made on the wall to hold coats. In its groove she found the bobby pin her father used to push in the button on the lock and open doors to her room and the living room. She took it, wishing she could take all his nails as well.

What did her parents know about love? Squeaky kisses in the kitchen, squeaky bed at night (perhaps she only imagined that). But her mother's stomach looked different. Perhaps there was another baby on the way. It seemed that what adults did was always right and validated by the Bible. Perhaps she should pity them all, especially the minister and his wife, who had to do something so shameful nine times just to go along with the Bible's command to be fruitful and multiply. Doing "it" would have been sinful if they hadn't been married. But it was difficult for Mandy to see people as pure and dignified after doing what they had to do. She wouldn't even let herself imagine them in the act. There was never any sign of where Pastor Palmer's babies came from because his wife was always confined at home until her babies were born. Then she came to church and thanked God for giving her a new baby as if it had dropped from the sky. It would probably seem no big deal: one person putting an organ into the body of another if people weren't so secretive about it—if everybody knew about it from when they were little and had gotten used to it.

She didn't really want to be angry with her father. She wasn't sure if her father was wrong, or if he was supposed to mind his daughter's business. But the next time he embarrassed her was no less painful.

It would have been a hot day, but the air was cooling on the water and running itself across Mandy's face where she lay in her bathing suit on a bank above the pond. Jerry came and sat on her legs and pinned her arms down. She'd dreamed of something like this happening. She looked up at him, hoping she didn't look too wide lying flat on the ground. It was the ideal place to get a kiss. Her lips were right below his, even if there was a foot of space between them.

She knew nothing until Jerry, who looked like the movie star Frankie Avalon, was jumping away. She followed the direction of his look and suddenly her father was yelling at her, "Get up and get your clothes on." She grabbed her towel and jumped to her feet. What had seemed natural—lying in the sun in a bathing suit —now seemed shameful and shocking. She ran across the hills, over the broken-down fence and down the path, storming into the house in tears. She passed her mother and ran into her room, locking the door. It was soon opened with a nail. Her father was telling her never to let a boy sit on her again. Shame *peased* through her. She felt dirty—and sorry that she lived in this house. Self-pity welled in her. She was accused of something worse than she had done, something unnamed. Jerry had done nothing except sit on her in the open, and among other people. She felt as if she was made of dirt. No odds, she reasoned, lifting her chin defiantly, if I do feel that way; everyone is made from dust—the same thing as dirt.

Her father liked "that fine young fellow" Philip, who sat in the front seat in the mission, clapping his hands when the congregation was singing "When The Saints Go Marching In." She wished she could tell her father about him reaching under her as she swam far out past the overfall, and how she felt she was going to drown as she struggled to get away from him. He finally gave up, laughing slyly.

Jerry hadn't done anything like that. But he avoided her after her father's reprimand. He was Church of England, and so it was just as well. She'd have to find a boy of her own kind, one who went to the mission. Philip's brother, Daniel, said he'd walk her home one night and she let him just to have the look of having a boyfriend. Perhaps she would let Daniel give her one kiss, though she wasn't sure she wanted to, but when she got to her door, she closed her eyes and pretended she was Marilyn Monroe. His lips touched hers; her lips relaxed. Then suddenly, to her horror and distaste, she felt a tongue against her teeth. She reacted quickly, clamping her teeth just in time. His tongue must have slipped back into his mouth, because he moved away from her. Pretending nothing had happened she fumbled with words to end the night.

"My father won't like it if he knows you kissed me good night," she mumbled.

He seemed to agree, already starting to walk backwards. He waved goodbye, then turned his back and walked away. She knew that this was the end of something that hadn't even happened. But the horror of what could have happened shamed her. The only tongue she wanted in her mouth was her own. She had heard about French kissing, but she didn't know if anyone but French people—the people who must have named it—actually did it. People in the romances she read were too refined for that, though young girls and fellows in *True Story* usually did, and they ended up in a lot of trouble.

She felt insulted, and shamed that he had tried to French kiss her. She couldn't tell anyone about it. Her friends would think she was some sort of—. She couldn't say the four-letter word. Her mind would feel even dirtier than her mouth was.

She'd found out about love from the Bible, but also from a Harlequin romance and *True Story* magazine that she and her friend had sent for from an ad in *Weekend Magazine*. They came to Janice's, but she'd sneaked them home to read, relieved to be giving them back to Janice before her father found them. She took a risk in bringing *True Story* magazines home. Hiding them under her mattress wasn't safe. They could disappear into her father's hands.

One day she was returning a *True Story* to Janice. She took it to school in a brown paper bag, and held it by her side as she went into the classroom. It must have been the look on her face that alerted Philip to the bag. He came towards her and grabbed it. She grabbed back and caught hold of it holding on from behind, not so much to retrieve the magazine as to keep her secret, as Philip, his long bony face filled with excitement, rushed to get it away from her. The bag slipped from her fingers as Philip, grasping it, fell over a chair in his way. Mandy fell on top of him, grabbing the bag in relief.

Mr. Rachit came in at that moment. His clammy white face, with black specks of newly shaven hair peppering it, suddenly turned red. His round cheeks seemed to shake and his eyes looked like two balls ready to hurl themselves through his spectacles as he thundered at Mandy, "I saw that bag sneaked in here."

Suddenly it was as if the teacher was reading the words in the magazine right off her face: the stories of sex and pregnancy. The story of Princess Margaret's and Captain Peter Townsend's love affair didn't count. She felt her cheeks heat up. Her eyes blazed at the thought that he was blaming her for Philip's nosiness. But she had her magazine. That was all that mattered.

She was glad she had gotten to know Tom Adrian. He wasn't just an art teacher who had rented her great-aunt's old house as a studio. He had asked her to call him Tom, and he was the only adult who was interested in discussing world events with her. He even brought up thoughts so far down inside her she didn't know she had them. She felt more alive, more alert, more aware of her value as a human being after she visited him. Her thoughts and her feelings always centred around concretes of her religion. He was helping her take abstracts and put them into shapes and colours without having to explain them in black and white. He made her feel as if she could suddenly see—as if she'd been blind before. She never imagined anything other than friendship; after all, he was ten years older than she was.

After school she decided to go see him. Perhaps she'd tell him about Mr. Rachit and the magazine. She lifted the latch and opened the gate and walked along a path to the house Tom called a studio. She'd always thought of a studio as something new and fancy rather than an old house that sat on the edge of a cliff with a 250-foot drop. As a little girl, she always felt lucky to be able to go in and come out without a piece of the cliff falling away and taking her with it, as had often happened in her dreams. She'd had so many dreams of stepping off the cliff and tumbling through the air—waking up startled.

She knocked.

"Come in," a heavy voice with an American accent called, and she pushed the door open and looked in around the doorway.

Tom was standing with his back to the door lifting a painting from an easel. He turned quickly towards her. "It's good," she said softly, catching her breath from having climbed the steep hill.

"Mandy, I've been thinking," he said, "sitting here painting and looking through the window." His lips half smiled and he turned fully to face her, "What a wonderful thing the first snowfall is. It is the epitome of peace in its purest." She followed his look to the window where a few snowflakes were beginning to tumble. People in the cove may have seen beauty in the seasons, but when they spoke of the coming winter, it was with the need to provide. When they spoke of summer it was with the expectancy of fish and an income—never a season for its own sake. He continued, "Winter is beautiful."

He went to check some fried potatoes he was warming on the stove. His head lifted towards her, "Have some." She shook her head, suddenly feeling as if she should not have stopped in. They moved inside and sat on the settee and he talked, looking into her eyes. His eyes always seemed to say something, but she didn't know what.

"I'd like to paint you without your clothes," he said softly. She wasn't startled or embarrassed. His interest came, she was sure, from being a professional wanting a subject to paint, and as clinical as a doctor wanting a patient to examine for diseases. Bodies were just objects that had a particular purpose.

She did not respond at first, and she wasn't flattered. Artists didn't paint pictures like she'd seen in a magazine in Jeff's room. Artists painted women with rolls where there were supposed to be hollows and dents, women who were all sizes and had all kinds of looks, beautiful and ugly.

"I'd rather be a girl Norman Rockwell painted," she answered swallowing hard. "I like clothes; I've been wearing them since I was born, and I can't change now." She knew she sounded silly.

"The only time Eve was perfect was when she wasn't wearing clothes."

Where had she heard that same reasoning? A strange feeling stirred deep inside her. Now she felt like disagreeing. "She wasn't too perfect," Mandy said flatly, "or she wouldn't have done what she wasn't supposed to do."

He came closer, tipped her chin up with his forefinger, and looked into her eyes. She turned quickly to avoid an intensity in his. They were like some unknown world that she'd dreamed about looking into, but she couldn't because she knew that once she did she would travel somewhere and not come back.

"Your virginity is tightly wrapped inside your mind as pure as God made it," he said quietly. "I like that."

167

She had read somewhere that when a woman gives herself to a man she can't get whatever she gave back. Did that mean he'd always have it, and what would he do with it after he got it? And what would that do to her? It was all so muddling. His warm lips brushed her cheek. "If you were in another country, would you have an affair with me?" he asked.

His question didn't seem to be real. It was as if it belonged to a villain in a Harlequin romance. She couldn't associate bad characters in Harlequin romances with real people, especially someone she was fond of. She suddenly felt so hot it was as if her face was squashed against the backside of the sun.

"I don't know," she answered, and wondered why she said that. To spare his feelings, she thought, thinking he would never really think of her in that way. She could never give all of herself for a little part of him.

She could hear the seas slamming against the rocks, and she felt a shiver. Her father hadn't objected to her going to St. John's with Tom and visiting the Art Gallery and having lunch at a Water Street restaurant. Her parents would have known if he wasn't to be trusted, wouldn't they? Perhaps it was no big deal. Her mother often said people should mind their own affairs and she knew she wasn't talking about sex, so perhaps he wasn't either. She tried not to think anything else about it. She pulled her collar up tight as the words she had heard somewhere echoed in her head: What you show, you owe.

"You're like winter waiting for spring. Don't let anyone ever spoil you," he said quietly.

"How can they?" she asked. "I know who I am, and what I want." He smiled as if he believed her.

His face suddenly darkened as he looked towards the radio in the kitchen. "I can't believe it!"

"What?" she asked.

"The president was shot today."

"Jackie Kennedy's husband!"

"Yes, John F. Kennedy. It's on the radio."

Mandy looked at him mesmerized. She thought of Jackie and her white mantilla and her perfect love. Now all she could see was a black veil. "I've got to go," she said abruptly.

"Please stay," he looked wounded.

"I can't."

His lean face was a lonely profile caught in the light of the window as she left the house. She closed the door behind her and crossed the path on the cliff to the sounds of rushing waves slapping against the cliff. The sky had cleared and her eyes lifted to pink clouds looking like burst balloons as she walked slowly down the road. She saw a neighbour watching her from the window. Perhaps she wasn't looking at her at all or thinking anything about her. Then slowly she began to acknowledge that the warm and gentle man she had trusted now had a stranger's face. When he had looked into hers she could only see him as a villain. She began to tremble and her trembling shook her. The friendship was over. She felt shamed. There was something black between them now.

Something had drawn her to Tom, but now it seemed as dangerous to her as the cliffs she crossed. They had created feelings between them out of a love for life. But that was as far as she could go with him. She liked being herself; she wanted to belong only to herself. She had to be careful from now on—to not let anything happen that shouldn't.

A memory of when she was ten flashed in front of her. She had been sitting up the side of a cliff with other kids. A snow plow was on the road beneath them cutting a path for the ambulance to take Cara Sue, who was fifteen, to the hospital. She could see Cara with her head lowered and leaning on her mother's arm as if she was as old as her grandmother. She looked as if her life had festered inside her body. Cara Sue had told people that the doctor who operated on her kidney had done something to her. That's why she was having a baby. Mandy had felt the cold under her bottom tunnel up to her heart, not because she was sitting on snow but because Cara Sue wasn't much older than she was, and if that could happen to her, it could also happen to Mandy if she did something she shouldn't, or if someone did something to her that he shouldn't. It was hard to straighten the facts of life out in her mind when she wasn't sure she had all the facts. Suddenly she was slipping off the crusty snow down towards the snow plow underneath. As she banged at the snow with her heel to made a dent that would hold her she almost pitched forward. A close call! How many of these would she have before she was old, she wondered. Sometimes after the Christmas story of the virgin birth she dreamed of seeing a Holy Ghost, one that told her she would have a baby whether she wanted one or not.

Her mind flashed back to the scene of Uncle Erlking, the old man she'd thought she killed. But that was another story, one she didn't want to think about right now.

As she walked down the dusty hill, she thought about Tom. She didn't want to get closer, but she knew from what was inside her that some day she would want to get close to somebody. There were times now when she became wary of an emotion stealing into her being that made her body want to be touched. If she closed her eyes, it was as if her flesh was already reaching out, even while her heart and her mind were closed off to it. Perhaps it wasn't wrong for Tom to want her. A door opened inside her. She felt a stir of pleasure before she slammed it shut. She was growing inside herself, thinking, creating a self she could be comfortable in, one whose judgement she could trust. She would never betray herself.

She hurried home and into her room. She pushed aside the wide lace curtains of her window and sat on the window sill until night came. The curtain fell against her hair like a white mantilla. She thought of Jackie Kennedy who must have lost a perfect love; she thought of herself hoping for one. She eyed the letter from Stephen on her bureau. A girl's name and strange address were in the left-hand corner. Who did he think he was kidding? Certainly not her father, who may have decided that Stephen was too far away to be a threat to her innocence.

She stared up at the moon looking like a lost ball in the lonely blackness of space. Sometimes it seemed that life was just as mysterious and as haunting—that love was as far out of reach. A cloud suddenly slid over the moon and when the moon emerged, it came as a silver claw.

Mandy let the curtain fall.

The truck heaved back and forth across the road with her father's hands gripping the wheel and trying to hold it steady. Mandy remembered her four brothers sitting in the open back of the truck. Suddenly the truck lurched and lifted, turning on its nose, tipping Mandy upside down. She heard the sound of gravel beating like hailstones against the cab, and her hand went up to grab the seat to keep herself from falling.

Suddenly her world ended.

THE ACCIDENT

Mandy had been feeling uneasy for months. A sense of fear had lodged inside her heart like a junk of cement and hovered over her thoughts like a black presence. Now in the cab of her father's truck, she pressed her foot against an imaginary brake every time her father drove too close to another vehicle on their way home from her cousin Sharon's wedding in St. John's. Her mother noticed her nervousness and asked what was wrong.

Unable to explain her feelings, she murmured, "Nothing!" She bent her head towards the bouquet of flowers she had caught at her cousin's wedding. They were real roses, all red and sweet-smelling.

When her father accelerated to pass a large transport truck, Mandy's feeling of apprehension became so intense that she let go of her flowers and cried out, "Don't pass it, Dad!" He ignored her outburst, and pulled out alongside the truck. He tried to pick up speed, but his truck was old and the driver of the transport truck didn't seem interested in letting her father pass, and then it was too late. They were going downhill and a car was coming towards them. As black as death! she thought.

She was going to die. There was no hope of her living beyond her fifteen years. They were all going to die, and as her mother tightened her arms around two-year-old Elizabeth, her voice was squeezed out of her, tight and coiled and ready to snap, "Eric, you're going to kill us all!" Words jumped to Mandy's tongue, and she wanted to speak them—to tell her mother not to distract her father, but all her life seemed to be frozen inside her as hard as a lump of coal. Her eyes felt glued to the bright blue sky, as if caught inside its dome. Fear had blocked out everything outside it, and nothing else existed. Her world was going to end here, at the pale blue line of sky her eyes were fastened to. If only she could write last words in black ink on its blank, empty blue, though she couldn't think what she would say. Then, like a miracle, her father swerved in time missing the black car by inches. Still there was no relief. The truck heaved back and forth across the road with her father's hands gripping the wheel and trying to hold it steady. Mandy remembered her four brothers sitting in the open back of the truck. Suddenly the truck lurched and lifted, turning on its nose, tipping Mandy upside down. She heard the sound of gravel beating like hailstones against the cab, and her hand went up to grab the seat to keep her from falling.

Suddenly her world ended. She was surprised to see it again from where she was kneeling, a long way, it seemed from the truck. She looked at her left hand hanging limp. The little knob in her wrist was missing, and she wondered where it had gone. She spat out gravel, forgetting her pain as she noticed Jeff sitting in a hollow under the truck, his head bent.

"He's dead." Her mother's voice came from a long way off. Mandy leaned back in horror. Her mother's open mouth was full of blood. Mandy had thought black was the colour of death. But red was everywhere. It was staining her beautiful white bag with the gold braided chain as it lay under Elizabeth's head. Elizabeth lay like a rag doll. Around her, petals of Mandy's bouquet of roses were scattered among gravel.

Suddenly Elizabeth screamed and her chubby little hands went to her head. Mandy wanted to get up and stop the blood running from her little sister's head. Taylor and Michael were sitting in bog looking at each other, blood smeared on their faces. Timmy was standing motionless, as if frozen still. Their father went from one to the other, blood running off the tip of his finger making a little trail on the ground. He called to Jeff, who, Mandy noticed, was trying to get out of the ditch under the truck.

Mandy felt relieved that no one was dead, but as she heard the wail of sirens she was afraid that there were invisible hurts that would kill them all. She sat back and began to cry, tears running down her face in hot streams, and dripping into the jagged cuts in her knees.

She didn't want anyone to lift her into the ambulance because her arm felt as if it was full of broken glass stabbing her in the bone. She closed her eyes tight, not wanting to see any more blood. Beside her, Elizabeth's cries blended with the sound of sirens. When Mandy opened her eyes again she had already been lifted into a bed at General Hospital.

Mandy lay in the white hospital bed feeling a rage she did not understand. Something had happened to her after the accident that made her feel as if she was someone other than herself. It was as if she had touched a strange face and had been frightened into a fierceness she could not understand. She tried to keep silent, but after awhile her agony became too much to hold inside. She found herself sobbing against the jagged cliffs of pain that her head had become.

She was glad that the rest of her family had gone home with only a few stitches in a different part of each one's body. She didn't mind staying where she was for awhile. The walls around her kept her safe after her terrifying ride, though none of the doctors seemed to know how to fix her smashed wrist, and they kept taking her to the operating room and bringing her back, leaving her with fire in her left wrist. There were nights and days of pain, and everything else was forgotten.

It was late one night when she heard a scream from the ward across the hall. She must have dreamed it, though it was almost as clear as the night the strange nurse came. Mandy had been drumming the fingers of her right hand along the guard rail and crying softly to herself. It had only been an hour since her medication but already the four-hour needle had worn off and pain was swirling her in a fog of misery. Suddenly she heard the rustle of a starched uniform. A nurse stood by her bed. When she spoke, Mandy couldn't pull her voice into answering. The nurse started to rub her hand along the back of Mandy's right hand. It was so soothing that she fell asleep almost immediately. She was surprised to open her eyes to a sunny morning. She didn't see that nurse again, and no one seemed to know who she was.

A few mornings later she heard a man scream. It wasn't her imagination, after all. Nurse Jeffers told her that the scream had come from a Russian on the male ward. The nurse seemed in the mood for chatting as she changed Mandy's bed. "Antanas was brought in late one night after falling down the hold of his ship as it lay anchored in St. John's Harbour. He broke both of his legs right off. Well, that's the story anyway," she said matter-of-factly. "Perhaps he was trying to defect. Who knows? I don't feel a bit sorry for him," the normally pleasant nurse added, her mouth tightening on her words. "If it wasn't for these people there would be no wars. You'd think he was honey the way some of the younger nurses swarm around him, washing his hair and doing anything else they imagine he wants them to do."

"I think I'll go have a look at him," said Mandy. Perhaps he was one of the Russians her father had called godless Communists. He had never met one, so he had to rely on what he read and heard about Russians. Some of them were persecuted Christians he felt sorry for. Mandy didn't know if she had ever seen a Russian before either, though she'd passed a lot of foreigners on Water Street. From the sounds of this one he was terrible, and probably a Communist.

Nurse Jeffers shook her head, "He's a surly fellow, for sure. He threw his bed pan after Mrs. Mossel gave him physiotherapy. She was only trying to move his legs for some exercise."

Mandy's heart stirred in pity. Here was a stranger in pain, who couldn't tell anyone what was hurting. "Pain can be understood in any language," Mandy said simply, determined she was going to see him, and give him the same kind of treatment the strange nurse gave her. It wasn't hard to imagine herself in a Russian hospital among strangers. Who would befriend her as she lay wondering what kind of information the Russian people had about Canadians?

She turned quickly as a bed was wheeled into the ward; a woman on it moaned softly. Mandy watched as she was lifted onto the bed beside hers. Tears slid out from under her closed lids.

Responding to Mandy's questioning look, the young hospital attendant told her that Mrs. Silvosky was a Polish lady—one of the last survivors of the plane crash at Gander that morning.

Mandy slid off her bed and went to hers. The woman looked old, not as old as Mandy's grandmother who was in her eighties, but older than Mandy's mother. Her grey hair was pulled back over her forehead and twisted into a knot at the back of her head. She had very high cheekbones, and deep shadows around the grey, deep-set eyes that suddenly opened. Mandy smiled at her and the woman's eyes seemed to grope for a friendly face as strange words snapped from between tightly clenched teeth. Mandy did not answer. She kept smiling at her, touching her hand. Mrs. Guinn, a little woman with a tanned, freckled face under a cloud of brown hair, pushed herself up in her bed and said she hated foreigners. "It was an immigrant," she said bitterly, "who took me from my little girl and husband. The idiot defected in Gander and was driving a vehicle around town shortly after, just long enough to run me off my legs, and here I am unable to move them and he's still on his two." She sank back on her pillow as if wounded.

The next afternoon, just as Mandy was waking from a nap, a nurse passed her bed with flowers for Mrs. Guinn who already had three arrangements. Her small face lit up. "Who in the world would be sending me roses?" she asked brightly. She buried her nose in them, then seeing a card she pulled it out. With a shriek, she hurled the bouquet to the floor. "That rat, how could he do this to me?" she screamed.

Mandy had been feeling envious. "I'll take them," she said breathlessly, rushing to pick the precious flowers off the floor. Their scent was not at all like the boy's love growing by her grandmother Maley's door—mixed with the pungent sting that the salty sea blew in over the land. They were pure perfume. She'd always wanted to grow roses, but flowers didn't grow on rocks and land sucked by the cold, briny sea, its slate cliffs, some grey and blue, others as white as the foaming waves that heaved against them with harsh slaps.

In Maley's Cove, summer was ushered in, not with the scent of trees and flowers bursting into bloom, but with the distinct scent of salt water spraying the air, and billowing over the outstretched arm of land that the sea continuously licked with a frothy mouth. "The brute who crippled me is trying to appease his conscience," Mrs. Guinn began to cry hysterically.

Suddenly disheartened, Mandy threw the flowers into the garbage can. Last year she had tried to grow poppies from a sample seed package she had ordered from the *Family Herald*. She planted them in ground her mother threw steeped tea leaves in. It was a moist spot. To her delight the poppies grew like coloured pieces of paper dancing in the wind, until a roaming cow bit them off their stems.

She could never ask her father and mother to spend money on flowers. They withered so soon that it would have been like asking them to burn the hard-earned money her father got from fishing. And it hadn't been a good year. Her parents bought her a new peach negligee with a white, chiffon rose pinned to the base of the neck. It had a nylon coat over it with long puffed sleeves. "Seeing that you'll be here awhile and not able to get dressed, it's good to have something nice to wear for visitors," her father explained gruffly, his weather-beaten face creased with worry, and his eyes carrying the fear that Mandy might lose her left hand. "And she a left-handed paddy too," she had heard him say.

She hadn't worn the coat yet because of the cast on her arm, but if she was ever going to see the Russian she'd have to rip the seams. She began to hook at them with a pair of manicure scissors Mrs. Guinn had given her. The sleeve soon slipped on over her cast. She lifted the hinged door on her table and looked into the mirror. Her blue eyes had lost the dull look she'd carried since the accident. And according to Mrs. Silvosky's Polish interpreter, who had a soft and gentle face, her hair wasn't red. "Red," he said, "makes one think of a bright Christmas colour. Human hair is never red—and it is never a solid color. Its many shades are brought out in the light. Your hair is full of autumn shades." Then it dawned on her that she had never had red hair. People who said she did just didn't know their colours. Her hair was auburn.

Mrs. Silvosky had been crying for awhile and Mandy thought she must be upset that she might not walk again. She went to her. "You," she said pointing to the Polish woman's legs, "you'll walk soon." She began tracking her fingers across the bedclothes to make sure she understood. She smiled and nodded, "Yak?" But when the Polish interpreter came back to visit her, he told Mandy that Mrs. Silvosky had believed the other patients were saying she was a Communist and she felt very sad that she could not tell them differently. Her holy cross was lost in the plane crash. Mandy found 'friend' in the Polish-English dictionary the interpreter brought and showed it to Mrs. Silvosky. "Yak!" she smiled, and reached to hug Mandy. The interpreter loaned Mandy his Russian-English dictionary. She checked some words and their pronunciations and decided she was going across the hall to see the Russian.

Mrs. Guinn offered her lipstick, and Mrs. Mckay dabbed perfume on her right wrist and even sprinkled some on her cast. Even Mrs. Silvosky gave her a smile of approval as if she understood what she was up to.

"The young are so energetic," said Mrs. Mckay in a tired voice, "not like us." The elderly woman's voice had a trace of envy.

"Speak for yourself," snapped Mrs. Guinn. "I'm not old and just because I'm laid out on this white sheet doesn't mean I'm dead."

Mrs. Mckay, ignoring the younger woman, said to Mandy, "Knock him dead, Kid."

Lifting herself on one elbow, Mrs. Guinn said testily, "She wants him alive, not dead."

"I don't want him at all," Mandy answered primly. "I just want to see him." She hurried into the hall, not wanting to be caught between the barbs of the two women.

She looked around to make sure there wasn't a nurse near, and then she went quickly into the men's ward. She found Antanas' bed in a far corner where she'd been told he was. He was lying motionless, his blond hair as straight as a whip, combed back over his head like strands of gold. He was wrapped in a cast up to his arm pits. She'd read somewhere that all Russians have to spend time in the army. Perhaps that was why he had such muscular arms and a broad chest. From where he lay flat on his back, his blue eyes looked out under long eyelashes. He looked just like the hero in the Harlequin romances she'd sneaked into her bedroom. She never thought she'd be part of a Harlequin romance— perhaps a whole chapter. From the first time since the accident she felt life filling her with a satisfying intensity she hoped would never leave.

"Spacibo," she said quietly, wondering if the pronunciation was correct.

The Russian turned and looked at her with eyes as cold—as cold as what? Siberia, that was it. Her father had read about Siberia in *Time* magazine and said it was cold, and people who believed in God were often sent there so that they would become atheists and get rid of unpatriotic beliefs. Now Antanas' silence was so strong she felt afraid of it.

"How do you feel?" she ventured in Russian. She'd practised it, and said it haltingly.

He understood and said, "Fine," in English. She stood by his bed for a few minutes even though he had turned his head back to the wall. She heard a rustle behind her.

"Now Mandy," Nurse Jeffers was saying patiently, "you shouldn't be on the men's ward unless you're a relative of one of the patients."

She shrugged in resignation. She couldn't really talk to someone who had his back to her, but she knew there would be other visits.

She heard him moaning later that night. She went out and in the dim light of the ward she saw him looking at the ceiling. He stopped moaning and turned to the wall when he saw her. She put her hand on his face and let her fingers caress his cheek. It was bold of her, but she saw him as a person in need, as she was when she was in pain and an angel of mercy had come to her. He turned and looked at her. She wasn't sure what she saw in his eyes, but he didn't look angry or suspicious. Nurse Jeffers came in with a needle. "Go," she said sternly.

"In awhile," she answered, not looking at her. She didn't care. She'd just found out from Mrs. Guinn that Antanas' physical therapist, who had called him a volcanic Communist because he screamed at her and threw a bedpan across the room, had been told to stop the Russian's therapy. New X-rays showed that there was a bone in his leg that hadn't set. It had pierced the skin, making any exercise excruciating.

The next time Mandy went to visit Antanas she had just reached his bedside when a short man, with white hair razed to his skull, came in. He had jaundiced-looking eyes, that appeared lidless, in a face that looked like sandpaper. At first he kept looking at her as if she had no right to be there. Finally he asked in a raspy voice, "Didn't your parents ever tell you not to talk to strangers?"

"No, but if you insist I won't talk to you," she said saucily. She had heard about him from the nurses. He was a Jewish store owner, fluent in several languages, including Russian. He was always complaining to the nurses about one thing or another. She would have liked to have talked to him since Jesus was a Jew and this man must be flesh of his flesh.

Instead, she decided to ignore the surly Mr. Steinberg as he tried to intercept her visits during the next few days. In the hall he hissed, "Are you trying to get that man killed?"

She turned, startled.

"He'll be killed when he goes home if you don't stop visiting him. He's being watched, you know, and your name is on a suspect's list," he added threateningly. He turned away, his pasty face a harsh mask.

"I don't believe you," she answered flatly, not feeling the least bit afraid. "I've never done anything to endanger Antanas." She moved past the man, more threatened by her doctor's news that she might need another operation on her badly damaged wrist, than by what he was saying.

The next time she went to see Antanas, he looked at her cautiously and she was afraid that Mr. Steinberg had told him lies about her. But then his hand reached towards the nylon white rose at her throat. She was afraid the back of his hand would slip to the hard knobs of her nipples, and she backed away, her face suddenly flushed. His hand, like a soft breeze, lifted to her hair and stirred it. She leaned towards him, then she looked around. Not seeing anyone, she bent her face towards his. At least he wasn't French; he wouldn't know about French kissing—or would he? Perhaps there was another name for the same thing. Her lips touched his. She drew back as if she was suffocating. When she moved away she saw him swallow hard. She wasn't sure what that meant.

She should have found out if he had a wife. What a terrible thing she may have done! She grabbed his Russian-English dictionary, and searched quickly for the word wife. "Zhenah?" she asked anxiously, hoping she'd pronounced it so he could understand.

"Nyet," he smiled. An amused look crinkled his face.

He picked up the dictionary and gave it to her. She looked at it and gave it back. He pressed it into her hand. She knew then that the dictionary would become the key to a dark door between them. Opened, each word inside would turn on lights of communication. Antanas' eyes brightened when she tried to speak whole sentences in Russian. He tried to say a sentence in English; she pretended she understood it.

Alison wasn't much older than Mandy. She and the other nurses knew that Antanas had a photo of himself. They had all asked for it. Alison said he had promised it to her. She washed his hair, and emptied the bag by his bed, and seemed to drool over him. She never helped Mandy wash her hair. "Why don't you cut that off?" she said sharply. "It's a mess." She'd snipe at her all day unless Mandy had just come from the operating room. Then she would ignore her.

Mandy had learned only the night before that Antanas was leaving the next morning. He was being sent back to a Russian hospital in Moscow still in his body cast. She should have listened to what he was trying to tell her instead of having to hear it from the nurses. Nurse Jeffers walked to her bed and told her that Antanas was forbidden to have visitors. He'd been cleared for his trip home. It was as if Antanas had been placed in quarantine, probably by the Jewish store keeper. She didn't even look to see if anyone was coming, but she hurried to him, and bent down to kiss him. She suddenly looked around to meet Mr. Steinberg's steely eyes. She turned her back on him and Antanas reached and pulled a strand of her hair down on his face.

She spent the night bent over her dictionary with intense concentration painstakingly putting together, from the puzzle of words, found in his dictionary, a last letter. Being left-handed, she hoped her letter, in the unfamiliar Russian alphabet, was not done too clumsily with her right hand. She finally finished the half-page letter and folded it over a picture of herself. She was so afraid someone would stop her from bringing it to Antanas that she couldn't sleep, and before the breakfast trolley came around she rushed over to the men's ward. She felt relieved that he was still there. She had imagined him disappearing in the middle of the night. Instead he was lying in his bed, his face freshly shaved and he was wearing a white shirt and tie. It was the first time she had seen him dressed in anything but hospital sheets. As he lay there with half-veiled eyes, she wasn't sure of her emotions. She knew that she had to give him her letter and picture before the officials showed up to take him away. She lifted his hand lying across his shirt and placed her gifts in his palm. He looked at the picture and letter and smiled, then he quickly lifted his shirt and stuffed the two gifts down his cast. Then reaching across to his dresser he picked up the coveted picture of himself and pressed it into her hand without a word or even a smile. His eyes were no longer cold; now they carried gold flecks in them, just like the sea on a sunny day.

Some memory slid in place and she wondered if this Russian was the stranger she thought he was. It may have been him she had seen the summer before wearing a cap and a short beard. He was dropping a strange coin into the photo booth at Woolworth's and she kept hearing a plop as it fell to the bottom of the slot. It looked like a fifty-cent piece. Instead, it was a strange coin with Joan of Arc on one side. "It's French!" she exclaimed. She'd thought then that the stranger must be French. Now she wondered if perhaps he'd picked up the coin in a French port. She had taken her quarter and given it to him. Then she'd picked up the stranger's coin. He had gestured for her to keep it, and now it was on a chain with a hole in it at home. It was a summer routine for Mandy to have her picture taken at Woolworth's on a trip to St. John's with her father. While he was at Leckie's getting building supplies, she'd walk along Water Street, and imagine it was a romantic setting for a novel. Then she'd go into Woolworth's and get her picture taken in her white angora bolero. A few of the pictures came out okay, but never as she imagined. Some pictures made her look like tar baby; others made her look like a ghost. The ghost ones had no freckles, which she was happy about. It was the same kind of picture she was giving Antanas.

She wasn't sure how she felt knowing that soon Russian officials would lift Antanas, still in a body cast, and take him away on a stretcher. Would she ever see him again? If only they could have gone to a park and enjoyed time in a normal setting! She started to walk away, then she looked back. She turned and went towards his bed. Leaning down, she wondered what he would do when she kissed him goodbye. His eyes looked into hers, liquid with tears. His lips touched hers lightly. Suddenly his hand came up from behind and cupped the back of her head, pressing her lips tight to his. He opened his mouth over hers. For a moment she couldn't move. There was such intensity in his kiss that she was afraid of it. She was relieved when he released her.

"Goodbye, Antanas," she said softly, then she pronounced it as well as she could in Russian, "Prashshai'te."

185

She backed away, and watched from a distance as Antanas was lifted onto a stretcher and wheeled into the open elevator where Russian officials, in dark overcoats and fur hats topping sombre faces, stood beside him.

He was going beyond her look, her voice, into a world she had read so much about, perhaps to be forever swallowed up in it, but he would carry part of her there in his memory. Tangible reminders of her would go all the way to Moscow. His picture would stay with her.

The elevator doors began closing, narrowing her vision of a Russian carrying the souvenirs of a romance her father could never have imagined her having, one he would never have allowed. As the doors clicked shut and Antanas disappeared, Mandy stood holding the little hand doll the Polish lady had made for her in the ethnic dress of the Polish people. She had travelled beyond her own world in a tangible way. She now knew that the Harlequin romances she had read were the stuff of real and imaginary people, living out their lives in unexpected twists and turns. She had started the first real romantic chapter of her own, written by life itself, and though it had ended almost at the beginning, she wasn't sad. Just maybe, another chapter could be written long distance. Perhaps Antanas would defect. Inside, she was dancing. The possibilities of love were endless. She could hardly wait to start the next chapter.

Glossary

backslidden — given up a commitment to one's religion

barbel — a bibbed apron made from flour bags and soaked in
linseed oil for waterproofing

beachy bird — sandpiper

beggar's claptrap — a noisy person who carries empty talk from one
place to another

bivvering — trembling or shivering

blaighard — vulgar, unbecoming language

britchins — codfish roe—orange-coloured and shaped like baggy
(bloomers) drawers

bulking — laying fish in piles, sprinkling each layer with salt

cantals — 112 pounds of codfish

carpenter — wood lice

clits — tangled knots in hair

coopy — comes from the word squat. It complements egg (coopy egg) because a child might squat to look for a hen's egg in confined spaces. Coop is the sound hens make. Some owners of hens used to call them hoops.

flat — small boat

frankum — a gum-like sap from the spruce tree

gentian violet — a violet-coloured dye used as an antiseptic

gowity (gowithy) — sheep laurel: low brushes with white and pink flowers

grainted — deeply imbedded with dirt

hyssop salts — a tonic, a stimulant

lunning — wind calming down

milquetoast — a light tan colour. The colour of tea with half milk

mimpsing — taking one's time drinking a cup of tea

ose egg — sea urchin

peased — seeped

piss-a-beds — dandelions

prating — talking

ragmoll — an untidy person or one who lets her clothes get ragged

rinding — stripping the skin off

scravel — rush about

slide — sled

sliered — looked slyly out from under half-closed lids

step-ins — skimpy underwear

streel — untidy person

strouters — heavy posts around the head of a wharf or stagehead

ten-test — a thick cardboard-like wall used in many homes

trotter bone — the bone of an animal—boen from the foot of a cow

vamp — a short woollen sock

widdershins — an old English word denoting a movement in a contrary direction to that of the sun—as of a clock whose hands are going backwards. (Brewer's *Dictionary of Phrase & Fable*. Centenary 1981 Edition, edited by Ivor H. Evans.)

wis — cry baby